VERCORS CAVES

GW00724657

VERCORS CAVES

Classic French Caving – Vol. 1

DES MARSHALL

Photographs by P. R. Deakin F.R.P.S.

Cordee – Great Britain

Copyright © Des Marshall 1993   Photographs P. R. Deakin 1993

ISBN 1 871890 71 3

British Library Cataloguing in Publication Data
A catalogue record for this book is available from the British Library

Frontispiece: Formations of the Salle Bouegin – Gouffre Berger

All trade enquiries to:
Cordee, 3a De Montfort Street, Leicester LE1 7HD

This book is available from all specialist equipment shops and
major booksellers. It can, along with all the maps mentioned
in the text be obtained direct from the publishers. Please
write for a copy of our comprehensive stocklist of outdoor
recreation and travel books/maps.

CORDEE
3a De Montfort Street, Leicester, Great Britain, LE1 7HD

Printed in Great Britain by BPCC Wheatons Ltd, Exeter

CONTENTS

ACKNOWLEDGEMENTS

I would like to thank the following for the assistance given in the preparation of this guide:-

'Edisud' the French publishers of the Speleo Sportive series of caving guides, for allowing me to use the maps and surveys from their publication 'Speleo Sportive dans le Vercors',

Donald Rust for transposing and upgrading the disk from my PCW onto a commercial disk,

Paul Deakin for allowing me to use some of his superb photographs,

John Gillett for writing and translating from French the map references section,

Ken Vickers who set up the contact with 'Edisud' and was enthusiastic about the production of the first volume and finally by no means least,

my companions on many trips. To them this work is dedicated.

Any faults in the guide are my own and as such I would welcome any updates or further information on the caves described or indeed description(s) of any cave you feel should go into a further edition. These can be sent to me; c/o Cordee, 3a De Montfort Street, Leicester, LE1 7HD.

PREFACE

Although this is an English guide to a selection of the caves in the Vercors region, I have maintained the French names for Chambers, Passages and Pitches because I feel that these should not be anglicised. I have also retained the word 'Siphon', for the same reason.

I have been as definitive and accurate as I can with the descriptions in the guide but errors may have crept in, despite my efforts. To avoid any errors creeping into future editions it would be appreciated if cavers send me, via Cordee, comments and further information on the trips here-in, or cave which you feel should be included. The trips described in the guide are to enable competent cavers to reach the bottom of the cave or to reach the normal finishing point for a trip, e.g. Grotte du Gournier as far as the 'Salle Gathier' or 'Salle Chevalier'. No attempt has been made to describe side passages or to describe anything other than the 'Voie Normale'.

I recommend that an extra rope or two be taken on the longer, more vertical trips - just in case. A 10m and 20m would suffice. It would also be politic to take a few extra hangers and slings.

INTRODUCTION

When one hears of the Vercors region, the cave that immediately springs to mind is the Gouffre Berger. Steeped in history – the first cave to go below the magical 1000m level – it is by no means the only cave in the area. There are many other caves which give magnificent descents and the idea of this selection is to help other cavers enjoy these really fine trips, either as part of a Berger expedition or as a separate holiday. The selection is purely a personal one, but all the caves described are regarded as some of the best the area has to offer. There is something for everyone, from experienced expedition cavers or for those embarking on their first caving trip abroad. The cave descriptions are not definitive as they only describe the 'voie normale'. For further information on the descriptions refer to either the 'Speleo Sportive' guide (in French) or 'Grottes et Scialets du Vercors' (again in French). Walks to the caves vary from 2 – 3 hours to less than 10 seconds!!

I have tried to simplify the method of finding the chosen cave by explaining the very complicated way that the French work out their map co-ordinates (map references). This may also be useful when referring to another area covered by the excellent 'Edisud' series of caving guides.

Covering a compact area of some 30km × 60km, the Vercors is a large limestone massif just south-west of Grenoble in the south-east corner of France. It is partly in two departmental areas – Isere to the north and Drome to the south. The area is a natural fortress being almost entirely surrounded by cliffs and there are two distinct levels: a lower one between 700m and 1200m where most of the towns, villages and agriculture is concentrated and a much wilder and barren plateau above 1500m with a high point of 2341m – Le Grand Veymont. The greater part of this plateau consists of N – S ridges and valleys. There are some particularly fine gorges that run E – W, notably those of the Bourne and Vernaison. Forestry is the main industry as much of the Vercors is covered with forests of pine, spruce and beech trees.

The Vercors, which became a National Park in 1970, is diverse in its fauna and flora. It is quite magical to come out of a hole at dusk, start driving down the road back to the campsite, to spot wild pigs with young darting away from the headlights. Above 1500m many flowers lend colour to meadows and bare limestone. Gentians, harebells and a host of other Alpine flowers abound here. Flowers along the hedgerows are a riot of colour in the summer months.

From the summits of 'La Grande Moucherolle' or 'Deux Soeurs' (Agathe and Sophie) or any other of the peaks along the Vercors central spine, magnificent views are to be seen eastward of the Dauphine Alps. Summits certainly deserve a visit especially if exploring the Antre des Damnes or Grotte des Deux Soeurs for example. Limestone pavements (lapiaz) here are also impressive, especially the one called 'La Purgatoire'!

I hope that this guide will encourage cavers who visit the area to appreciate it and to gain as much pleasure as I have over the years exploring caves where there are no bureaucratic problems.

Situation

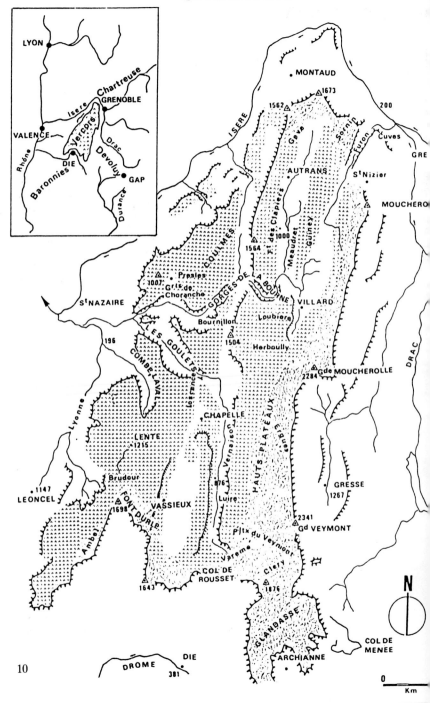

WHERE TO STAY

There are many camp sites in the region. Some have more facilities than others! For the Gouffre Berger for instance the preferred camp site for many is on the Sornin Plateau at La Moliere. However, there are no facilities there, so if you, or your family, prefer to have running water and toilets on site with a swimming pool within easy reach then there are camp sites at Autrans and Meaudre with some excellent places to eat in the evening. Both sites make ideal centres for exploring the caves that are in the north of the region. They are some 20-25 minutes away from the plateau – which can sometimes be shrouded in cloud when the sun is shining down at Autrans or Meaudre.

More sites exist but my favourite place to camp is at La Balme de Rencurel in the Bourne Gorge. There are two sites here which are much cheaper than those at Autrans or Meaudre and have full facilities including a nearby river for washing ropes and gear after a muddy trip! There is a good cheap hotel in La Balme de Rencurel (£12 per night per room, summer 1992) in case it rains and you get flooded out of your tent. The municipal camp site on the Auberville side of Pont en Royans is often crowded and can be noisy, but is handy for some excellent restaurants in the town. Villard de Lans has camp sites, but these are more expensive than staying in the Bourne Gorge.

If concentrating on the caves to the south of the region then La Chapelle en Vercors may be the preferred centre. Good sites are available here with a swimming pool in the town.

WEATHER

Generally, during the summer the weather is good in this area. However, as in all mountainous areas the weather is susceptible to sudden change. Here, as in the Alps spectacular thunderstorms occur which manifest themselves as flash floods underground. The return journey from the bottom of a deep (or even not quite so deep) cave can be vastly different from when you entered. Always rig as far out of the water as possible. The drain off areas for many of the caves is huge, allowing enormous quantities of water to enter the caves. One has only to visit the Goule Blanche in the Bourne Gorge after a period of heavy rain to be suitably impressed with the volume of water surging out of its entrance!

If you can understand French there is a very good recorded forecast, which is updated every 6 hours to be obtained by telephoning 76 51 11 11. If your French is non-existent then try a local tourist office. There is a good one in Villard de Lans by the main car park (free) and a small one in Pont-en-Royans as well as in Autrans.

GEAR

THE place to visit is EXPE where you can buy anything to do with caving and climbing. The owner, Georges Marbach, is a caver himself and speaks English, as do some of the staff who are all extremely helpful. Carbide (Carbure in French) is readily obtainable here and is much cheaper than in Britain. The more you buy the cheaper it is! (·5 francs per kilo if you buy a 70 kilo tin, otherwise it is 13 francs a kilo – 1992 prices). This is a good place to stock up. Unfortunately the shop closes in August. The address is EXPE, BP 5, 38680 Pont-en-Royans, France. Telephone 76 36 02 67. They take Access and Visa as well as hard earned cash!

EXPE is found by taking the road towards Romans-en-Royans for about 4 kilometres to a left fork signposted to Auberville. About 2 kilometres further on EXPE is on the right, an isolated building looking much more like a warehouse than a shop. In fact, equipment especially undersuits and PVC oversuits are also manufactured here.

It is also possible to purchase caving equipment in Grenoble but the shops are much more difficult to find. As in Grenoble, Villard de Lans also has outdoor equipment shops but the prices are somewhat more expensive than at EXPE.

Occasionally it may be possible to obtain 'carbure' from hardware or agricultural stores, but this is a less sure source.

If you use electric lamps you will have to take a charger complete with adaptor and then pay for an electric point at the camp site (if it has one!). Far easier and better to use a combined carbide/electric unit. The flame from the carbide provides essential warmth in the event of an enforced stay underground and if enough carbide is taken unlimited light and warmth is given. A segment of old car inner tube is indispensable for carrying spare carbide. (Make sure there are no holes in the tube first!!).

CONSERVATION

As with any other caves worldwide, conserving their fragile environment is of paramount importance. All too often spent carbide litters the floor of caves. It is a very simple matter to take spent carbide out of a cave. All that is needed is a small mat to shake the spent carbide from the generator onto so that the powder can be poured into a container like a BDH. At the present rate of dumping spent carbide thoughtlessly, the caves will be an unsightly mess before too long. In some caves hundreds of little grey piles spoil dry, beautiful passages, e.g. Gournier and Berger. Cavers do not cave to see piles of carbide, they cave to see their beauty. Many a time cavers remark "what a bloody mess!" SO PLEASE BRING YOUR SPENT CARBIDE OUT and dispose in a dustbin or rubbish bag.

Formations too need protection. Careless or thoughtless manoeuvres can spoil in a fraction of a second a formation that has taken thousands of years to form. Mud encrusted hands spoil formations too, as pressure ingrains the mud into the whiteness and purity of the formation. It is possible that "holding on" can snap a formation leaving an ugly stump behind. Be careful, in other words, what you hold onto.

If you take food and drink with you on a long trip underground please bring empty tins (they really are lighter on the way out), wrappers and food out and dispose in the dustbin or rubbish bag at the camp site. Please also respect the mountains. It may be that you will spend a night at P2 or a couple of nights at the Antre des Damnes and although bivouacking is tolerated in the mountains please ensure that ALL your rubbish is brought back down. Animals live in the mountains and a thoughtlessly discarded tin lid, plastic bottle or broken glass bottle can seriously injure and maim a creature.

Remember – take nothing but pictures and leave nothing but footprints!

MAPS

Since June 1992 the French have started a series of maps to popular areas and retitled them TOP 25. Although still part of the Serie Bleue system, these popular maps are now known by the new title and of course cover a larger area. The "TOP 25" MAPS required for finding the caves in this book are as follows:

Carte IGN 1/25,000 No 3235 OT Autrans/Gorge de la Bourne/Parc Naturel.

Carte IGN 1/25,000 No 3236 OT Villard de Lans/Mont Aiguille/Parc Naturel.

Carte IGN 1/25,000 No 3136 ET Combe Laval/Foret de Lente/Parc Naturel.

The Carte Touristique Maps are now no longer available. However, some of you may have these maps and the ones for the area are:

Carte IGN 1/25,000 No 226 Vercors Nord. Carte IGN 1/25,000 No 227 Hauts Plateaux Nord. Carte IGN 1/25,000 No 228 Hauts Plateaux Sud.

Anyone having the old 'Serie Bleue' maps will still be able to use the following without having to buy the new TOP 25 series. The ones to use are:

Carte IGN 1/25,000 No 3135 Est, St-Marcellin.

Carte IGN 1/25,000 No 3136 Est, Combe Laval/Foret de Lente.

Carte IGN 1/25,000 No 3234 Ouest, Tuillins.

Carte IGN 1/25,000 No 3235 Ouest, Villard-de-Lans.

Carte IGN 1/25,000 No 3236 Ouest, La Chapelle-en-Vercors.

All the Cave References remain the same whether 'Serie Bleue' or TOP 25.

A good all round map of the whole area is the 'Massif du Vercors' (Number 12) in the 'Editions Didier & Richard' on a scale of 1/50,000. It has some of the caves marked on it, as well as being overprinted by the GR (Grande Randonee) paths, ski-ing and equestrian tracks. These maps also show huts and include a schematic diagram showing the time it takes to walk from hut to hut.

The new TOP 25 series maps and the 'Massif du Vercors' No 12 maps are available through Cordee, on application.

HOW TO DETERMINE MAP REFERENCES ON FRENCH MAPS

1 – References of a point on the map

The idea of references comes from the following. We live in a three dimensional space and if we want to indicate a point A on the surface of the Earth we have to use three independent measurements.

It is convenient to use three directions at right angles as a reference, starting from the same point O, the origin of the references. The projection of A on each of these axes, made parallel to the two others, allows us to measure the 'References of A' = X, Y and Z.

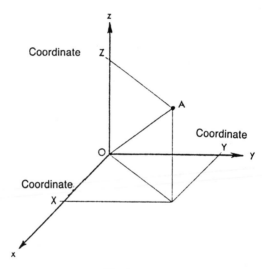

Fig. 1

2 – Lambert References

Using the chosen system of reference, the origin O and three axes Ox, Oy, Oz, there is an infinite number of ways of indicating the point A. We want to chose the most convenient reference directions possible.

One that stands out is the vertical direction, which is simply its altitude. The axis Oz is the vertical axis.

We now need to chose two directions in the horizontal plane at right angles to each other. It is simple to chose Lambert North Oy and its East perpendicular Ox. To improve the ease of use of the system a grid of kilometre squares is constructed parallel to Ox and Oy covering the whole of France.

16

On the IGN maps of scale 1/25,000 the intersections of the grid are shown as small crosses every kilometre.

On the IGN maps of scale 1/50,000 the intersections are only shown every 5 kilometres.

The origin of the Lambert Grid system is situated in the South West of France at sea level. The Y references cross the map from South to North, and the X references from West to East. The start of the grid system and the integral values of the references are given on the borders of maps.

3 – Determining the references of a cave entrance

This is a simple operation by using a ruler on the map.

Pencil in the kilometre square that surrounds the cave entrance on the map. To determine the entrance X reference, measure the distance in millimetres from the left hand side of the square to the entrance, e.g. in the diagram below this is 31mm.

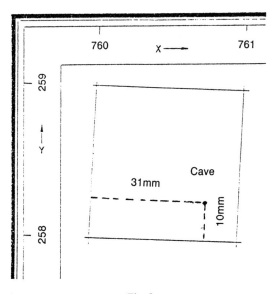

Fig. 2

Remembering that for a 1/25,000 scale map, 1mm on the map represents 25,000 millimetres (25 metres) on the ground, you can calculate that this corresponds to 31 × 25 = 775 metres. The left hand side of the square corresponds to the integral value of the reference 760. X = 760,775.

For Y you do the same, turning the ruler through 90 degrees and measuring the distance from the entrance to the bottom side of the square. If this measurement is 10mm, then Y = 258,250 (10 × 25 = 250).

17

Be careful to read the references in the direction that the numbers increase. A common mistake is to read them the other way round (e.g. X = 761,225) but it is easy to avoid this with a little bit of care.

As for the altitude Z, it is read from the map by counting the contour lines.

The use of a 'Perspex' graduated map reading square allows you to avoid such mistakes. These are easily obtained from map shops in the area.

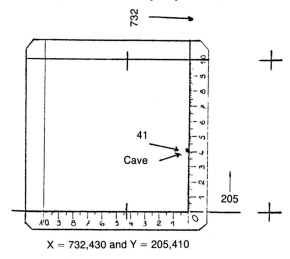

X = 732,430 and Y = 205,410

Fig. 3

With a sharp pencil, mark in the line of the kilometre square surrounding the entrance, immediately BELOW the entrance on the map, starting from the left hand edge, to get an accurate base line. Place the Perspex square on the pencilled line and slide it to the right until the vertical scale touches the point to be marked. Read off directly: X = 732,430   Y = 205,410.

4 – Pinpointing a cave on the map from its Lambert References

This is the reverse of the operation just described.

To pinpoint the cave with references: X = 659,350   Y = 224,700   and Z = 1240m.

On the map, mark two axes 659 (Vertical) and 660 (Horizontal). They meet at one of the grid crosses, from which you can pencil in the square defined by the axes 224 and 225 on one side and 659 and 660 on the other.

You now have to measure vertically for Y, (700 divided by 25 = 28mm) and horizontally for X, (350 divided by 25 = 14mm) into the interior of the square on the map to pinpoint the cave entrance. A 1mm spread on each axis corresponds to a square of 25m × 25m on the ground, which is good enough to go straight to the entrance you are looking for (In theory!!).

You should take care to check that the altitude Z corresponds to that on the map, after working out your X and Y, by checking the contour lines, as this allows you to detect any mistakes in working out your X and Y!

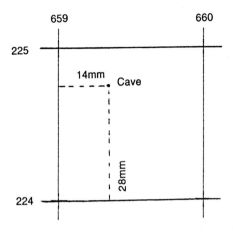

RESCUE

It is to be hoped that no-one has an accident underground but in the unfortunate event that one does occur you can perhaps evacuate the victim yourselves if the injuries are not too severe and if there are enough of you in the group. More severe injuries will need specialist stretchers especially if a broken back or neck is suspected. However, many of the caves, although sometimes long and deep, are explorable by two fit and competent cavers. Although four cavers is the ideal number – one to stay with the victim and two to go out for help – some may prefer to cave as a pair which is fine, until something goes wrong.

If you are not able to organise your own rescue team or if the injuries are too severe for your resources then recourse must be made to the local authorities. If caving in the north of the region, Isere, (e.g. Berger, Glaciere d'Autrans, Trou qui Souffle, Gournier, Grotte de Bury etc.), then the number to telephone is 76 36 01 32. In the south, Drome, (e.g. Glaciere de Carri, Grotte des Ramats, etc.) then the number to call is 75 48 22 38. If in doubt call the local Gendarmerie who will then contact the right people.

Remember that Cave Rescue is not free as it is in the UK so make sure you have sufficient insurance BEFORE leaving home. Ensure too that you will be covered for helicopter evacuation because some of the caves are very remote and far from a road (e.g. P2, Antre des Damnes, Scialet de la Combe de Fer, Grotte des Deux Soeurs, etc.). The BCRA Insurance scheme is probably the best around and is reasonably priced. Contact BCRA 20 Woodlands Avenue, Westonzoyland, Bridgewater, Somerset TA7 0LQ.

DEPTH AND LENGTH OF CAVES

DEPTH OF CAVES
- Gouffre Berger .... −1278m
- Antre des Damnes............................ − 723m
- Grotte de Gournier .... + 680m
- Scialet de la Combe de Fer .... − 582m
- Grotte de Bury .... − 520m
- Grotte de la Luire.............................. 513m (−450m, +63m)
- Trou qui Souffle............................... 400m (−345m, +55m)
- Grotte des deux Soeurs .... 332m (−315m, +17m)
- P2 .... − 319m
- Reseau Christian Gathier....................... 309m (−107m, +192m)
- Scialet de Trisou .... − 273m
- Scialet de Malaterre .... − 230m
- Scialet de l'Appel .... + 199m (although following
           the water downhill the measurement is taken from the resurgence at the
           Grotte du Brudour)
- Glaciere de Carri.............................. − 196m
- Glaciere d'Autrans............................ − 180m
- Gour Fumant .... − 163m
- Grotte Favot .... − 118m
- Pot de Loup .... − 94m
- Grotte des Ramats .... 86m (+59m, −27m)
- Grotte de Bournillon .... − 65m

LENGTH OF CAVES
- Trou qui Souffle/Les Saints de Glace.................... 33,169m
- Gouffre Berger..................................... 22,400m
- Grotte de la Luire 17,799m
- Grotte de Gournier 15,125m
- Reseau Christian Gathier 9,406m
- Grotte de Bournillon.............................. 5,950m
- Scialet de l'Appel................................. 5,900m
- Grotte de Bury.................................... 4,900m
- Scialet de la Combe de Fer 3,400m
- Grotte des deux Soeurs 3,250m
- Grotte de Ramats.................................. 2,500m
- Antre des Damnes................................. 2,500m
- Gour Fumant 2,203m
- Scialet de Malaterre............................... 1,600m
- Scialet de Trisou.................................. 1,388m
- Glaciere d'Autrans 1,300m
- Grotte Favot 850m
- Glaciere de Carri 380m

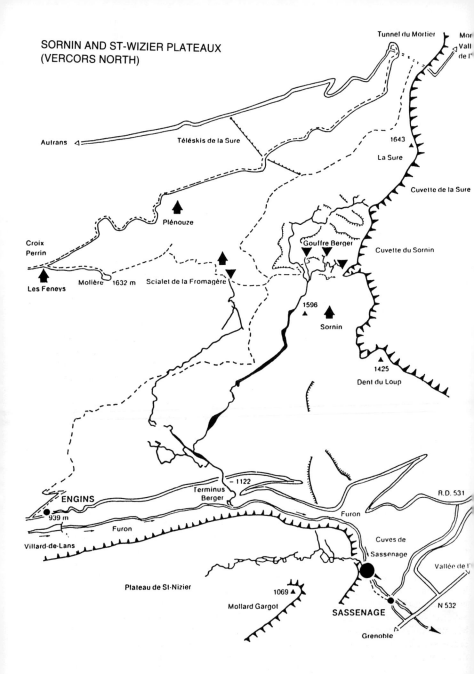

SORNIN AND ST-WIZIER PLATEAUX
(VERCORS NORTH)

Tunnel du Mortier

Mor
Vall
de l'

Autrans

Téléskis de la Sure

1643
La Sure

Cuvette de la Sure

Plénouze

Croix
Perrin

Gouffre Berger

Cuvette du Sornin

Les Feneys

Molière   1632 m   Scialet de la Fromagère

1596

Sornin

1425

Dent du Loup

1122

ENGINS

Terminus
Berger

R.D. 531

939 m

Furon

Furon

Cuves de
Sassenage

Villard-de-Lans

Vallée de l'

Plateau de St-Nizier

1069

Mollard Gargot

SASSENAGE

N 532

Grenoble

22

GOUFFRE BERGER

Map: Carte IGN 1/25,000 No 3235 OT Autrans/Gorge de la Bourne/Parc Naturel (TOP 25). The cave is marked on the map.

Map Reference: X: 856,63   Y: 329,44   Z: 1460

Depth: −1278m
Length: 22,400m

There are 3 other entrances to the Berger – Puits Marry, Gouffre des Elfes and the Scialet des Rhododendrons.

Access
This is the only cave in the guide where formal permission has to be obtained. It is advisable to write at least 18 months in advance if not 2 years! The address is: Monsieur le Maire d'Engins, Mairie d'Engins, 38360 Sassenage, France.

On arrival in Autrans take the road to la Moliere, where many expeditions camp, which is found by taking the D218 towards the Tunnel du Mortier. Before you reach the tunnel keep going to the right and continue to les Feneys where by going to the left you reach the car park at the Col du Moliere. DO NOT LEAVE your valuables in your car! To find the cave leave la Moliere by walking in a northerly direction along a well marked track. After an animal enclosure take the first turning to the right. From the descending path neve is often seen in the Scialet des Ecritures to the right. Follow the path marked with intermittent blue paint marks and cairns past the Scialet de la Fromagere (Scialet d'Engins), which is situated under a huge spruce tree. Continuing downhill a depression is found where the GR9 crosses east to west and there is a grassy area that has been nicknamed 'Camp des Anglais'. At the next path junction turn 90 degrees to the right. After 350m keep left and follow cairns across the lapiaz. The path descends many small steps and eventually the entrance to the cave opens out in a depression. It takes about 45 minutes to reach the hole but when you are tired after a long trip inside it may take over an hour to walk back to the campsite. It is advisable to mark the way back with fluorescent strips for ease of navigation in the dark. It could be misty, as well as dark even in day time!

History
The Berger was found during a reconnaissance trip on the 24th May 1953 by J. Berger, Bouvet, Ruiz de Arcaute and Jouffrey. Later the team reached the top of 'Puits Garby' at −103m and turned back because they had no ladders! On the 13th and 14th July, Berger was accompanied by Cadoux, Gontard, Garby, Arnaud, Marry, Potie, Brunel, Lavigne and Sillanoli and descended all the entrance series of pitches to reach the 'Grand Galerie' at −250m and continued to 'Lake Cadoux' where they stopped. A 28 hour round trip. In November 1953 another expedition was mounted and strengthened with the help of F. Petzl after his successes in Dent de Crolles. 'Lake Cadoux' was passed and the 'Tyrolienne' was reached at −372m. Another long trip of 38 hours. Meanwhile Petite Didier placed some dye in the water at 'Petit General' (Little General) which showed in the Cuves de Sassenage some 48 hours later. It was not until the summer of 1954 that the next expedition took place comprising 13 members including, this time, P. Chevalier. The 'Grand Eboulis' (Big Rubble Heap), the 'Salle de

Treize' (Hall of Thirteen, named after the explorers), were discovered. They continued to find the river below 'Salle Germain' reaching and descending 'Claudine's Cascade', −712m. A total trip time of 142 hours!!

In November 1954 the bottom of the 'Grand Canyon' was reached at −903m. By now the trips were of Himalayan proportions as many months of preparation were needed to obtain equipment and people had to make special leave arrangements from work. During the summer of 1955 an expedition lasting 218 hours reached the top of 'l'Our-agan' (Hurricane) at −985m. At the same time the 'Galerie de la Boue' was explored to −230m. In August 1956, during an International meet, −1000m was passed for the first time in a cave anywhere in the world. The siphon at −1122 was reached during a 380 hour trip!! This same expedition also found the 'Puits Marry' and made a junction with the Berger.

In 1963 during a British expedition Ken Pearce dived the first siphon. 4 years later, helped by the Pegasus Club, he passed the second siphon. At the same time and in the years 1967/68/69 the Speleo Club de la Seine and the F.L.T. explored the 'Affluent −1000' and the 'Reseau de l'Ouragan' (−955m). Between 1973 and 1977 a club from the Rhone-Alpes continued explorations in the 'Galerie de la Boue'. In 1975 the Gouffre des Elfes was found and the S.C. de Vizille made a junction with the Berger. In September 1978 F. Poggia, P. Penez and F. Vergier dived siphons 3 – 5 and reached −1148m.

In July 1978 B. Faure of the S.G.C.A.F. resumed work on a cave known as S1 on La Sure, Scialet des Rhododendrons (previously discovered in 1971 by the F.L.T.). After many visits to enlarge the passages they reached the river and a siphon at −250m. F. Vergier and F. Poggia passed the siphon on the first dive to emerge in the Berger. In July 1982 P. Penez dived siphon 5 and reached −1248m. In 1991 the Scialet de la Fromagere was connected to the Gouffre Berger by a 205m dive giving a total depth of −1278m.

Description

The 15m entrance pitch quickly leads to the 'Puits Ruiz' (27m) at the top of which is a rotting wooden platform put there during the original explorations. This leads straight into the 'Holiday Slides' which consist of three separate drops and on to the 'Puits du Cairn' (35m). Water can be found up the side passage. From the bottom of 'Cairn' shaft the meanders begin. Although not tight, they are gloomy. They have occasional stem-ples placed across helping progress but each year seems to see the demise of one or two more. 'Puits Garby' (38m), splits the meanders into two. 'Puits Gontard' (28m) is at the end. There are some long drops in the floor and some awkward climbs here. 'Puits Aldo' (42m) is soon reached. This is a fine shaft but care has to be taken to ensure that the rope is hung well out to the right (facing out). Water comes in....! At the bottom a couple of small climbs (spare short ropes useful here for coming out) lead into a large passage from under a huge boulder. This emerges into the 'Grand Galerie' with the 'Galerie Petzl' entering shortly after from the left. Further down, 'la Riviere sans Etoiles' (the Starless River), soon reaches 'Lake Cadoux' which is often dry but can become impass-able without a dinghy. Leaving a dinghy and a 60m pull back cord on the down side will ensure that you can pass this on the way out. Heading on down, a superb array of stalagmites are seen in the 'Salle Bourgin'. Pass between stal to reach 'Petite General' (10m) followed by the 'Cascade de la Tyrolienne' where the huge 'Grand Eboulis' (Big Rubble Heap) enters. The best way through the huge house-sized boulders is difficult to find but by heading over to the right you should reach Camp 1. (One year I saw a pram complete with wheels among the boulders!). The 'Salle de Treize' (Hall of Thirteen) is a magnificent sight. 'Balcony' pitch precedes the 'Calcite Slopes', (climbs both up and down). 'Vestibule' (15m) is in two sections entered through an arch. Traverse lines reach the top of a short vertical bit (best abseiled into the streamway below) and the start of the 'Coufinades' (Canals). Traverse wires help to keep one out of the deepest water to reach 'Cascade Abelle' (5m). This is rigged well clear of the water. The character of the cave changes here by becoming active for the first time. By climbing, traversing and wading you reach 'Cascade Claudine' (17m), at the bottom of which is a large pool. Downstream passed the 'Cascade des Topographes' (5m) enters the 'Grand Canyon', a long bouldery descent in a huge gallery. The stream has been swallowed far below and the cave's character changed to one of quietness. At the bottom of the gallery Camp 2 is found among the jumble of boulders. 'Puits Gache' (20m) is descended soon after leaving Camp 2. The route is dry at first leading to the water again. Two smaller pitches, 'Resseaut Mat' (10m) and 'Ressaut Singe' (10m) quickly follow each otl.er. 'Grand Cascade' (27m) comes almost immediately after 'Ressaut Singe'. 'La Baignoire' (4m) and 'Vire tu Oses' (Little Monkey) follow. Little Monkey is easily recognised by the amount of tackle usually left draped across the wall. A traverse across the right wall leads to a drop and traverse to the head of 'l'Ouragan' (Hurricane). This 44m drop is aptly named as it is very windy. There is a camp among the boulders away from the bottom of the pitch and away from the draught, but is not a particularly good place unless you have no alternative. The steady downward trend of the cave passes the 'Affluent −1000' (1000m inlet) on the right. A scramble over boulders, down short cascades and deep pools leads to a short climb up into a dry oxbow, bypassing a small pitch. Rejoining the stream again, more wading and traversing over deeper pools leads to the 'Pseudo-siphon'. This is too long to traverse and wearing a wet suit is advisable for the swim. However, there is usually a motley collection of, hopefully, inflatable boats to reach the true siphon.

Equipment

Shaft	Depth	Rope	Fixings	Remarks
Entrance	15m	10m	2 bolts	Free climb 6m
Puits Ruiz	27m	40m	3 bolts	Wood platform
Holiday Slides	15m	15m	6 bolts	Three separate drops
Puits du Cairn	35m	40m	8 bolts	
Puits Garby	38m	55m	6 bolts	
Puits Gontard	28m	40m	8 bolts	Traverse to pitch head
Relay pitches	20m	35m	6 bolts	Three drops of 5/10/5m
Puits Aldo	42m	60m	7 bolts	Traverse to pitch head
Petit General	10m	20m	3 bolts	Ladder useful here
Cascade de la Tyrolienne	5m	27m	2 bolts	12m for pitch and 15m for traverse after pitch
Balcony	15m	40m	4 bolts	
Calcite slopes	25m	42m	2 bolts	12m rope for climb up and 30m down. Thread belay for climb
Vestibule	15m	25m		Natural belays
Cascade Abelle	5m	15m	3 bolts	
Cascades	10m	20m	5 bolts	Free climable in low water
Cascade Claudine ..	17m	40m	4 bolts	
Cascade des Topographes	5m	25m	3 bolts	
Grand Canyon		15m		Useful for handline
Puits Gache	20m	25m	4 bolts	
Resseaut du Mat ...	10m	20m	3 bolts	
Resseaut du Singe ..	10m	30m	3 bolts	
Grand Cascade	27m	45m	4 bolts	Natural belay
La Baignoire	4m	10m	2 bolts	
Vire tu Oses	45m	50m	6 bolts	
Ressaut de l'Ouragan	10m	20m	2 bolts	
Puits de l'Ouragan ..	44m	60m	6 bolts	Traverse to clear water

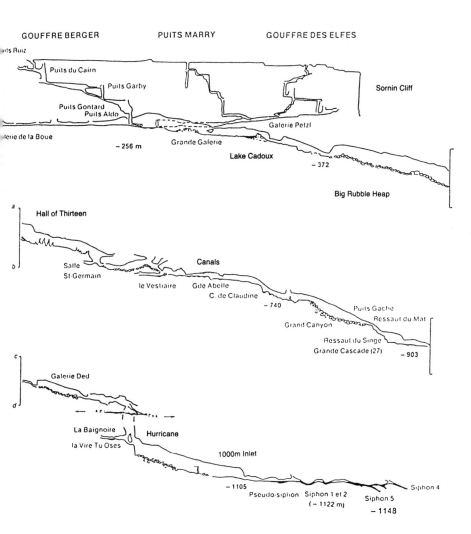

GOUFFRE BERGER   PUITS MARRY   GOUFFRE DES ELFES

uts Ruiz

Puits du Cairn

Puits Garby

Sornin Cliff

Puits Gontard
Puits Aldo

Galerie Petzl

lerie de la Boue

– 256 m   Grande Galerie

Lake Cadoux

– 372

Big Rubble Heap

a

Hall of Thirteen

b

Salle   Canals
St-Germain

le Vestiaire   Gde Abeille
           C. de Claudine

– 740   Puits Gache
              Ressaut du Mat
      Grand Canyon

      Ressaut du Singe
c   Grande Cascade (27)   – 903

Galerie Ded

d

La Baignoire   Hurricane
la Vire Tu Oses

1000m Inlet

– 1105   Siphon 4
       Pseudo-siphon Siphon 1 et 2
              (– 1122 m)   Siphon 5
                           – 1148

GOUFFRE BERGER

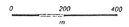

0   200   400
          m

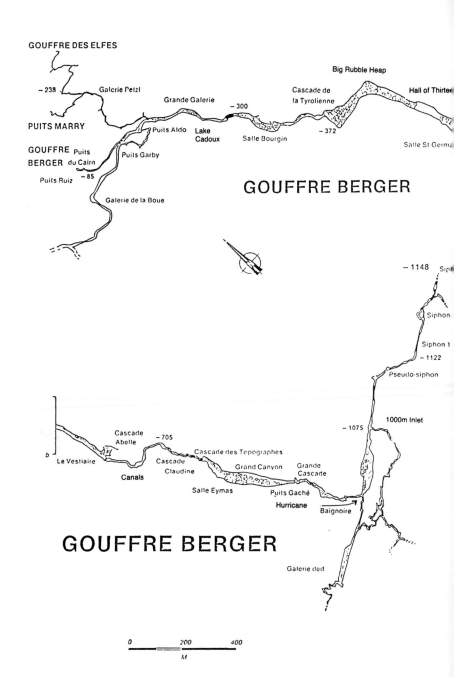

GOUFFRE DES ELFES

− 238

Galerie Petzl

Grande Galerie

Big Rubble Heap

− 300

Cascade de
la Tyrolienne

Hall of Thirtee

PUITS MARRY

Puits Aldo

GOUFFRE
BERGER

Puits
du Cairn

Lake
Cadoux

Salle Bourgin

− 372

Salle St Germa

Puits Garby

Puits Ruiz

− 85

GOUFFRE BERGER

Galerie de la Boue

− 1148

Sip

Siphon

Siphon 1

− 1122

Pseudo-siphon

1000m Inlet

Cascade
Abelle

− 705

− 1075

b

Le Vestiaire

Cascade des Topographes

Cascade
Claudine

Grand Canyon

Grande
Cascade

Canals

Salle Eymas

Puits Gaché

Hurricane

Baignoire

GOUFFRE BERGER

Galerie ded

0   200   400

M

28

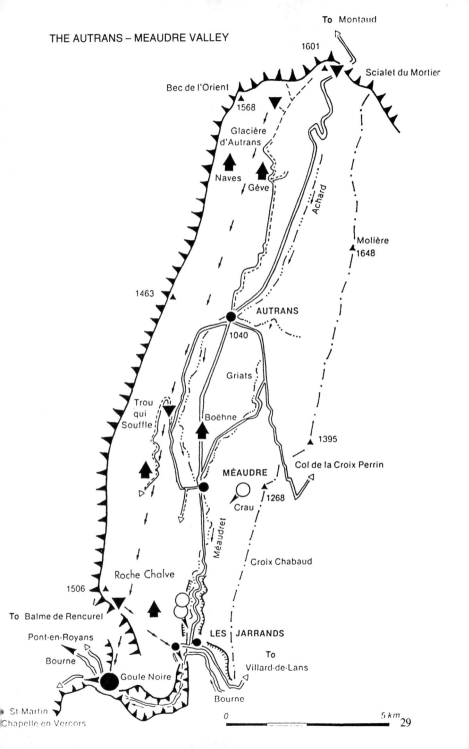

THE AUTRANS – MEAUDRE VALLEY

To Montaud

1601

Scialet du Mortier

Bec de l'Orient

1568

Glacière
d'Autrans

Naves

Gève

Achard

Molière
1648

1463

AUTRANS

1040

Griats

Trou
qui
Souffle

Boëhne

1395

MÉAUDRE

Col de la Croix Perrin

Crau

1268

Méaudret

Croix Chabaud

Roche Chalve

1506

To Balme de Rencurel

LES JARRANDS

Pont-en-Royans

To
Villard-de-Lans

Bourne

Goule Noire

Bourne

St-Martin
Chapelle-en-Vercors

0   5 km

29

GLACIERE D'AUTRANS (OR GLACIERE DU PAS DE LA CLE)

Map: Carte IGN 1/25,000 No 3235 OT Autrans/Gorge de la Bourne/Parc Naturel (TOP 25).

Map Reference: X: 853,12   Y: 330,96   Z: 1398

Depth: −180m
Length: 1300m

Access

From Autrans take the narrow road to the Cabane de Geve then follow the 'Route Forestiere du Cyclone', a forestry track, to a large clearing where you can leave your car. Follow the very rough track (GR16) uphill out of the clearing until a vague path is seen on the left. This is easily located as it is some 100 metres beyond a stone painted white with the number 367 etched into it painted in red. Follow this path, past La Fontaine on the right, until it leads steeply down into a large depression, generally having ice in the bottom of it, about 35 minutes from the car park in the clearing. The obvious entrance to the right of the depression is the 'Glaciere' whilst the other entrance the 'Patinoire' is found up to the left. A slippery descent takes you to the entrances.

History

In August 1936 the Speleo Club de Paris explored the first shafts but stopped at −45m. In the summer of 1963 the Speleo club de Saint-Etienne and the G.R.E.S.S. reached −110m at the start of the meanders. The following year they reached −170m. In 1965 an accident stopped explorations for some years. In 1972 the S.G.C.A.F. reached the bottom of the cave at −180m. During the years 1973-74 they discovered an ascending passage which was followed to −13m.

Description

The ample entrance leads immediately in to a slippery meander which is best dealt with on a rope as the meander steepens the nearer one gets to the pitch head. It is not known as a 'Toboggan' for nothing!

The first pitch (Puits Englace) is 32m and the amount of ice here varies. One year I had to climb up 10 feet before descending the far side and down the pitch. Another time the slope of ice led straight on to the pitch. A number of bolts of dubious vintage exist but with care a safe free hang can be arranged for this superb pitch. Care is needed on the final 6m due to an overhung area. The landing is amongst ice blocks. P8 starts almost immediately. Take care with loose stones. P11 follows and leads straight onto P27 immediately followed by P24. These last two pitches are tremendous but can be quite dangerous when wet. Even a small amount of water will make them most unpleasant. A small hole at the base of P24 leads down P6 and onto the final pitch before the meanders.

The character of the cave now changes dramatically. From being in spacious shafts the cave shrinks into a quite tight and very rough meander. Shortly before a squeeze a series of short shafts, P9, P6 and P8, lead to the bottom at −180m.

Equipment

Shaft	Depth	Rope	Fixings
Toboggan		20m	3 bolts
Puits Englacé	32m	45m	4 bolts
P8	8m	11m	2 bolts
P11	11m	25m	3 bolts
P27	27m	35m	3 bolts
P24, P6 and P5		55m	7 bolts
P9, P6 and P8		45m	3 bolts and natural belays

Most people will be content to reach either the base of the shafts at the start of the meanders or to reach the bottom of the cave at −180m. However, for those of an exploratory nature it is possible to ascend the shafts that lead up to −13m by climbing P25 (30m rope), P15 (20m rope), P19 (25m rope) and P15 (20m rope). There are various bolts and natural belays. Beware of stonefalls!

The 'Patinoire' is worth a look into as there is a frozen lake (Salle de Glace), beyond the entrance pitch P4. Another pitch, P6, leads to a 'T' junction where, by going right you overlook the 'Puits Englace' of the 'Glaciere'. Carrying on down very rough meanders covered in fine coral like stal leads to two more pitches P4, P8 and the end of the cave.

Equipment

Shaft	Depth	Rope	Fixings
P4	4m	10m	2 bolts
P6	6m	10m	2 bolts
P4	4m	6m	1 bolt
P8	8m	12m	2 bolts

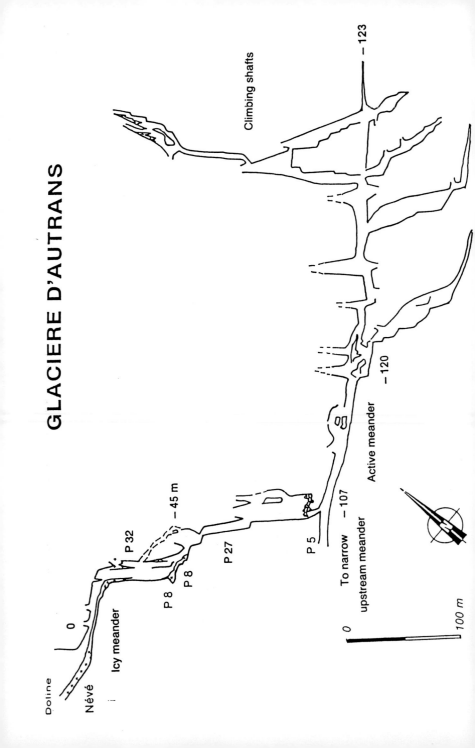

GLACIERE D'AUTRANS

Doline

Névé

Icy meander

0

P 32

P 8

P 8

P 27

− 45 m

VI

P 5

− 107

To narrow

upstream meander

Active meander

− 120

Climbing shafts

− 123

0

100 m

RELATIONSHIP OF THE GLACIERE D'AUTRANS WITH THE PATINOIRE D'AUTRANS

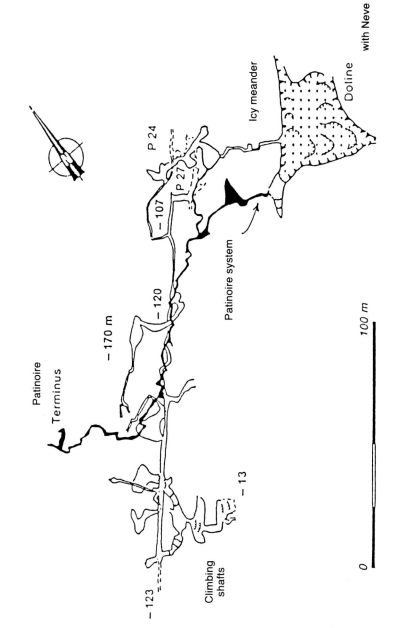

Patinoire
Terminus

− 170 m

− 120

− 107

P 24

P 27

Icy meander

Patinoire system

Doline

with Neve

− 123

Climbing shafts

− 13

0   100 m

TROU QUI SOUFFLE and LES SAINTS DE GLACE

Map: Carte IGN 1/25,000 No 3235 OT Autrans/Gorge de la Bourne/Parc Nature (TOP 25).

Map Reference: Trou qui Souffle   X: 850,05   Y: 321,51   Z: 1072

Depth: 400m (+55m, −345m)
Length: 33,169m

Access
From the village centre of Meaudre take the old road to Autrans past the ski complex in Meaudre to the hamlet of Cochet. 600m beyond here the road splits. The right hand turn follows the valley to Autrans but you need to go straight on, climbing gradually.

2km from Cochet on the left hand side of the road just after a bend the cave entrance, which is an extremely draughty and cold slot, is located. There is ample parking on the opposite side of the road. It is possible to place the first bolts while sitting in the car seat!

To find Les Saints de Glace continue up the road to the sharp left hand bend. From the apex of this bend a rough forestry track leads off right and it is possible to park 270m along this track where a well marked path descends to the cave into the wood on the right, (an 80m walk).

History
The entrance to the Trou qui Souffle was opened by the construction of the forestry road in 1909. The first explorations were carried out by Andre Bourgin in 1936. On the 21st December 1939 he made a short trip as far as the 30m shaft. On the 18th January 1940 he reached the 'Salle de la Vire'. Because of wintry conditions and the war Bourgin did not return until the 29th August 1941, with his friends Abry, Fontgalland, Raymond Gache, Joutet Manchon, his wife and 3 youngsters from the Chantiers de Jeunesse. They reached −165m in an expedition lasting 13 hours.

In the same year the 9th group of Chantiers a Meaudre explored in detail the entrance series making a circular route close to the entrance, finding the start of 'Reseau des Araignees' and the 'Galerie Excoffier' to the level of the 'Salle de la Vire'. On 30th August 1942 Bourgin and Raymond Gache reached −186m and inscribed their names on the wall. At the time it was the deepest cave in the Vercors.

In 1953 the Clan des Tritons de Lyon found the 'Galerie des Condensations' leading to the 'Puits Cigale', the 'Riviere de la Toussaint' and the 'Reseau Pont d'Arc'. In 1955 the same clan dyed the siphon at −208m. This took 48 hours to show at the resurgence of Goule Noire. During the period 1962-64 the Group des Cyclopes explored the 'Reseaux Cigale' and 'Gault' (+60m). They also found a short cut in the 'Reseau Pont d'Arc' and found the start of the 'Meandre Francois'. In January 1974 the F. L. T. re-discovered the upstream system of 'Pont d'Arc' as well as the window giving access to the 'Meaudre Francois' and continued their explorations. In 1977 the G. S. M. de Fontaine re-explored the 'Reseau Francois'. The following years they explored adjacent passages and the 'Reseau Tonton' to −250m. In March 1980 B. Leger dived the siphon at −220m then the Societe Hydrokarst explored 300m of new passages. In June 1980 the S.G.C.A.F. (E. Pouard and B. Lismonde) passed the choke at the north end of the 'Galerie Francois' to gain a junction with the 'Salle Hydrokarst'. The same month B. Talour and F. Leclerc (S.G.C.A.F.) discovered a passage beneath the 'Galerie Francois'

which came back under the 'Conciergerie'. In March 1981 the choke at the end of this lower passage was passed by J.J. Delannoy and B. Lismonde giving access to new passages, 'Galerie de Paques', 'Paques Sud' and reaching a low point of −313m. Since then a number of small passages have been explored, including the 'Reseau Vivam', a passage upstream of the Toussaint and the Saints de Glace, (S.G.C.A.F.). in 1985 B. Leger dived the siphon at −208m and made a junction with the siphon at −220m.

It had long been thought that there must be a second entrance to the Trou qui Souffle. Explorations from the inside had been terminated some 7m from the surface which had been determined by molefone and on October 15th 1989 after improvements 'Les Saints de Glace', the second entrance, was born.

Description

The Trou qui Souffle has two distinct rock types and can be split into lower and upper systems which are joined in a number of places. Three trips are described in the upper or "Senonien" limestone and a fourth – the through trip – links the two "Senonien" entrances with the lower or "Urgonien" limestones.

1: To siphon Verna −220m. The entrance pitch is 7m dropping into a small chamber. Take the main route out of this into a canyon passage which leads to the fine P30 immediately followed by P8 and two free climbs of 3m and 7m (handline useful). The passage slopes steeply down to the 'Salle de Vire'. A fixed handline to the right (of dubious vintage) can be followed or a 4m descent in the full flow of any water. Following the handline around to the right a large block needs to be descended. This separates one from the drop on the left. At the end of the sloping ledge a 6m descent reaches the floor of the chamber. Go down the slope to the left to rejoin the stream to where a canyon passage leads to P12 quickly followed by another P12 at the bottom of which is the Galerie des Condensations' and on to P6. This leads via a short climb into the Bourgin Streamway too tight at water level after a short distance. A short traverse allows the streamway to be regained just before it disappears. Follow the high level fossil passage to the 'Salle a Manger' (ignoring the low level passage leading off the right) and then take the higher level phreatic tube on the left. Soon after the tube becomes more rifty. A roof tube on the right (straight on leads to the end of the Bourgin streamway in 30m) passes a cross rift and becomes a hands and knees crawl. After 30m a hole in the floor leads to the 'Puits Noye' which is by-passed by continuing for a further 20m to a second hole giving an easy free climb down. Right leads back to the pitch while left leads to 2m and 4m climbs (very slippery). These lead to a keyhole shaped streamway, passable at roof level. After 30m the streamway widens and a 9m climb regains stream level. The stream disappears down a small hole but a dry passage leads off above. Follow the left hand tube down into a large steeply sloping passage and the sump at −220m.

2: The 'Reseau Cigale'. Follow 1 above to the bottom of the second P12 and take the Galerie des Condensations' marked by a strong draught (in summer). The fairly large passage with some deep pools reaches a crossroads. To the right a window allows the Puits Cigale' to be seen and the noise of the cascade to be heard. To the left a small muddy passage is followed. This passage continues to the 'Reseau Toussaint' and Reseau Pont d'Arc' but this is not the way. Going up between these two passages a small meander is entered. This bypasses the 'Puits Cigale' and 20m above there is a junction. To the left is the access to the lower or "Urgonien" system (15km of passages) but to the right a calcite slope descends to the head of the 28m 'Puits Cigale'. A 30m rope is used for the start of the descent of the shaft. Part way down you meet a stream issuing from a meander with the water cascading down the rest of the pitch. The meander is entered and is perhaps the finest in the system being ·5m – 1m wide and 5m – 15m high. There are small flooded holes in the floor and there is fine scalloping on the walls. On coming out of the meander enter a very wide but not very high passage composed of heavily corroded and brittle rock. This is the level of the 'Reseau Gault'. The rock here has a greeny tint to it and marks the boundary between the "Senonien" and "Urgonien". The passage has scattered blocks fallen from the roof. 200m further on a 6m cascade is climbed on the left (handline in place). Before entering spacious gallery it is necessary to crawl in knee deep water for a short distance. The passage gets large and larger to a point where huge blocks covered in a thin layer of calcite are strewn across the floor of a chamber. Height is rapidly gained and after an area of powdery rock, a superb meander is entered and followed to its end. The surface is not far away at this point!

3: Les Saints de Glace (As far as the 'Salle Hydrokarst'). This second entrance gives perhaps the easiest way of reaching the "Urgonien" or lower passages. The pitches are small with the highlight of the trip being the descent into the 'Salle Hydrokarst' to reach the siphon at −269m. For approach see Access above (page 34). Water is close by. Once through the enlarged entrance tunnel two small crawls enter a chamber. At the far side of this chamber the first pitch is met – P11. There are some ropes of dubious use in place. The chamber at the base of this pitch has some large blocks and the second pitch – P9 – quickly follows. The pitch head is behind a very large block on the right. At the base of the pitch a rift passage leads off and is interspersed by small chambers. The rock tends to be rough with lots of chert sticking out. Three small pitches follow – P4, P6, P3. The stream disappears, swallowed by a small hole. The rift narrows and there are three narrower bits (23cms). On entering a larger passage the water returns. A little further on the water permanently disappears and perhaps reappears in the 'Galerie Francois Nord'. The rest of the trip is in fossil passages and starts off in an ample meander, closer to the horizontal, soon reaching a 'Toboggan' −176m. This is the key to the return trip. There is an arrow to indicate the way out!! Descending the 'Toboggan' a phreatic tube is seen on the left heavily scalloped leading to the 'Siphon Hydrokarst'. The way on is to the right and soon reaches the balcony overlooking the 'Salle Hydrokarst'. This chamber is the largest in the Trou qui Souffle. An 11m pitch reaches the floor of the chamber at −226m. The chamber marks the junction between the "Senonien" and "Urgonien" limestones. Descending the chamber northwards to the siphon at −269m huge scallops are to be seen on the ceiling. (Some are as long as 2m!).

4: Trou qui Souffle – Les Saints de Glace Through trip via the 'Galerie Francois'. This is the finest trip. It is very interesting as it incorporates the best the system has to offer "Senonien" passages, an "Urgonien" meander and an abandoned fossil passage not to mention the two largest chambers in the system – the 'Salle Hydrokarst' and the 'Conciergerie'. The trip is ideally done by two teams going in opposite directions as finding the way is quite complex and unless you are sure of the way pull through trips could leave you stranded!! Entering by the Trou qui Souffle go as far as the 'Galerie de Condensation' and pass the 'Puits Cigale' on the left. Continue in the passage that leads to the end of the 'Reseau Pont d'Arc' (see 2 above). This low passage ('Laminoir de Bassine') can have water 8cms – 12cms deep which can be lowered by siphoning. (There is a pipe in place). The rest of the crawl is very muddy. Once out of this horrible crawl traverse to the left around a shaft (line in place) to arrive above the short 50m 'Galerie de l'Anastomose'. At floor level a small window allows another deeper meander to be reached. Do not take the crawl 10m before this! There are a number of shafts and climbs in this meander which eventually ends in a larger, sloping passage. There is a large choke to the left which is the far side of the choke in the 'Reseau Tonton'. A 6m pitch descends into the lower or "Urgonien" system. A shower of water descends into this passage – 'la Cascade Tonton'. Descend to overlook the 'Conciergerie' by the superb 'Meandre Francois' with 6 short climbs. The 'Conciergerie' is entered by a spectacular 15 pitch. Keeping up to the left an arrow indicates the way into the 'Reseau Sud', 'Cuspide' and the 'Galerie Pacques'. Going down and across to the left hand wall of the 'Conciergerie' meet the 'Galerie Francois' which starts at the 'Pas du Loup', a 1·5m drop (rope for the small!). (Along the large passage a turn to the right can be taken – the 'Galerie des Marmites' – and although a fine passage it is quite aquatic.) Continuing along a beautiful passage on the right hand side, ignoring a left turn, climb up 5m (rope in place) to reach a passage with many holes (marmites) in the floor. 100m further on a short climb up leads to a passage that ends in a choke. Climb 5m up to emerge in the 'Salle du Soupirail' and on into the 'Salle Hydrokarst'. Exit as for Les Saints de Glace (page 36).

Equipment

Shaft	Rope	Fixings	Remarks
Trou qui Souffle to −220			
Entrance shaft	10m	2 bolts	
P30	55m	6 bolts	High traverse out
P8	15m	3 bolts	
Salle de Vire	10m	2 krabs	Fixed ladder and rope
P12			
P12	35m	6 bolts	Combine the two
P6	10m	2 bolts	
Reseau Cigale			
Puits Cigale access to meander	30m	3 bolts	Pendule
Cascade	5m	pitons	
Cascade Cigale	10m	3 bolts	Ledge and climb
Reseau Francois			
Traverse over Pont d'Arc. . .	18m	4 bolts	Hand line
P6	8m	2 bolts	
Traverse	6m	2 bolts	
Traverse	6m	2 bolts	
P5	8m	2 bolts	
P7	10m	2 bolts	
P7	10m	2 bolts	
P8	10m	2 bolts	
Traverse	6m	2 bolts	
P7	10m	2 bolts	
P5	8m	2 bolts	
Conciergerie	20m	3 bolts	
P5	7m	1 bolt + natural	

Equipment

Les Saints de Glace

Shaft	Rope	Fixings	Remarks
P11	25m	4 bolts	Traverse to pitch head
P9	22m	Natural 3 bolts	
P4	10m	2 bolts	Traverse 3m to pitch head
P6	13m	3 bolts	Traverse 3m to pitch head
P3	5m	2 bolts	
P5	10m	2 bolts	
P11	20m	3 bolts	Traverse 2m

A survey and other trips can be found in the excellent book 'Le Trou qui Souffle' by Baudoin Lismonde. Although the text is written in French the surveys are excellent. See Bibliography.

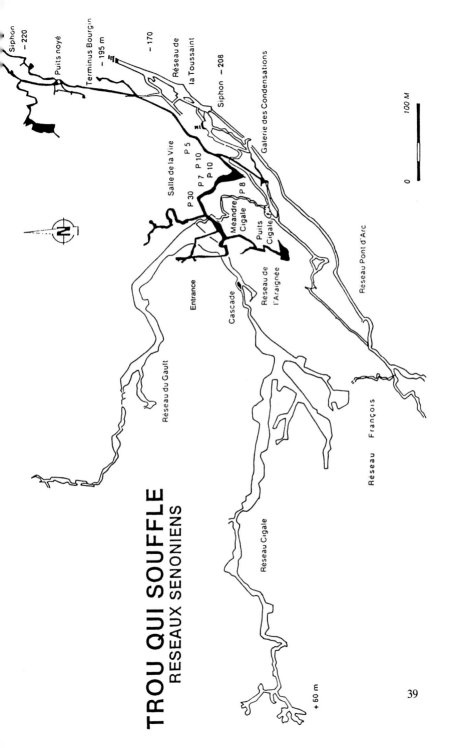

TROU QUI SOUFFLE
RESEAUX SENONIENS

Siphon − 220

Puits noyé

Terminus Bourgin

− 195 m

− 170

Réseau de la Toussaint

Siphon − 208

Galerie des Condensations

Salle de la Vire

P 5

P 7 P 10

P 10

P 30

P 8

Méandre Cigale

Puits Cigale

Réseau Pont d'Arc

Entrance

Cascade

Réseau de l'Araignée

Réseau du Gault

Réseau François

Réseau Cigale

+ 60 m

100 M

0

N

39

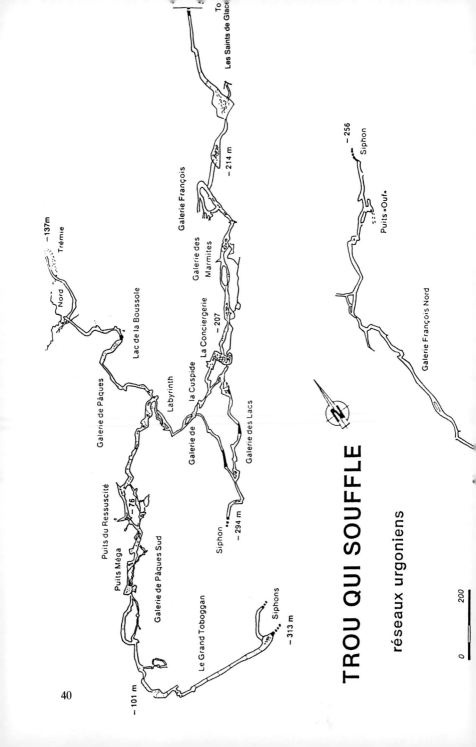

TROU QUI SOUFFLE

réseaux urgoniens

- 137m
Trémie
Nord
Galerie de Pâques
Lac de la Boussole
Labyrinth
Galerie de
la Cuspide
Galerie François
Galerie des Marmites
La Conciergerie
- 207
Galerie des Lacs
Puits du Ressuscité
- 76
Puits Méga
Galerie de Pâques Sud
Le Grand Toboggan
Siphon
- 294 m
Siphons
- 313 m
- 101 m
- 214 m
To Les Saints de Glace
- 256
Siphon
Puits «Ouf»
Galerie François Nord

0   200

40

THE COULMES MASSIF AND THE CHORANDE CIRQUE.
VERCORS WEST

To Malleval

Grotte de Bury

1398 m
Pas de l'Ane

To
Romeyere

Mont Noir

▲ 1475

To St-Pierre
de Chérennes

1421

● Le Fa

Combe de
Ravassière

Col du Mont Noir

1373 ▲

Refuge du Serre du Sâtre

Fontaine
de Pélouze

To Rencurel

863 ●
Presles

Charmeil

Grotte
de Gournier

Le Grand Serre 1297 m

To
Villard de Lans

To Choranche

Rochers de
Presles

874 ▲

Grotte de
Coufin

La Goulandière
1096 ▲

To
ont-en-Royans

Grotte
de Chevaline

Choranche

Pre Martin

La Bourne

265 m

Source Odier

La Bourne

0   100 m

Barrage de Bournillon

41

GROTTE DE GOURNIER

Map: Carte IGN 1/25,000 No 3136 ET Combe Laval/Foret de Lente/Parc Naturel (TOP 25). The cave is marked on the map.

Map Reference: X: 840,78   Y: 313,04   Z: 580

Depth: +680m
Length: 15,150m

Access
The cave is situated in the Bourne Gorge close to the Choranche show cave. From the road that runs through the Gorge take the signposted road leading to the show cave and park car in the main car park. It is important to ask permission (readily granted) from the owner of the show cave to visit the Gournier. Follow the path through the souvenir shops and cafe to the show cave from where a signposted path leads left. A junction is soon reached. Left leads down to the 'Cascade de Gournier', but carry straight on to reach, after a short descent, the stream issuing from the cave. Turning right here the large entrance and lake is soon reached. 20 minutes from the car park.

History
Long known, it was not until 1889 that Decombaz explored the entrance lake. In 1947 J. Deudon crossed the lake and entered the fossil gallery. This was explored in the same year to +166 metre, Terminus Bourgin (J. Deudon, A. Bourgin, Penelon and Sage). Two years later the same team found an access to the river beneath and followed it to a 12 metre high waterfall. This waterfall was climbed by P. Chevalier, L. Eymas, A. Sillanoli in 1952 but stopped at the 'Salle Chevalier' at +200 metres.

In 1960 the S.C. Seine (A. and G. Marbach) and the F.L.T. explored the 'Affluent des Parisiens' and followed the main river to the first siphon, the Siphon Jerome and also found the upper passages of 'Deux Jeans' an unexpected passage off the 'Allee Blanche'.

In 1973 the S.C. de Lyon, M. Bugnet, R. Chenevrier and P. Licheron passed the siphon Jerome but were stopped 350 metres further on at a second siphon (+300 metres). This was passed the following year by P. Licheron. On the 2nd June 1974 M. Bugnet and R. Chenevrier followed the passage for 500 metres. In November 1975 the S.C. de Lyon were stopped at +460 metres and 5·3 kilometres from the entrance by a 5 metre cascade.

In November 1976 R. Chenevrier, M. Schmidt and D. Trouillex tried to pass this obstacle but were swept away by a violent flood thus stopping further explorations for a number of years. In 1980 the S. C. de Lyon went to the second siphon. On the 2nd May 1981 the S. C. de Dijon (P. Degouve, B. Lebihan and J. Michel) passed the 5 metre cascade and explored 2200 metres of new passage to arrive (at +605 metres) at a boulder choke. In February 1982 the same club passed the boulders to explore a further 1370 metres of new passage but were stopped at +645 metres. A new expedition, in July 1982 reached the present limit of the cave, (+680 metres).

Description
Using the boat cross the lake to a landing on the left after 30 metres. A 6 metre climb aided by a metal plate leads to an obvious 30 metre rightward traverse. This is followed

to emerge into the huge fossil gallery at some large gours which are sometimes filled with water. The large (10 metres × 20 metres), spectacular and beautiful passage continues through the 'Allee Blanche' and after 700 metres the 'Salle des Fontaines' is reached. The first access to the river is 150 metres further on. Descend carefully a funnel shaped depression, with black arrows pointing down, through boulders to the river.

A better descent is the 2nd access which is found 250 metres beyond the 'Cascade Rouge'. This access, found on the right of the passage, is a short, narrow rift coated with calcite. It leads to a climb down boulders onto a ledge overlooking the river, from where an easy traverse leads into the river just beyond the boulder choke crawled through on the way from the first access. It is, however, a little more difficult to find than the first access. If descending to the river through the first of these, a short, easy streamway leads to the boulder choke which can be quite nasty in high water condition. It is awkward to pass this, though not particularly tight. As the second access enters the streamway immediately beyond the choke it is easier to recognise on the way back.

There are two other accesses to the river beyond the two described, but they need equipment.

The superb streamway continues past many cascades and wire traverses above deep pools to the 12 metre cascade which which needs to be climbed with care. A rope is usually in place but take your own, just in case! 100 metres further on the large 'Salle Chevalier' is reached. This is the usual point for returning to the entrance but it is possible to ascend a 40 metre pitch from the 'Salle Gathier' to reach the 'Affluent des Parisiens' and the passage leading to the Siphon Jerome at +267 metres.

WARNING: The streamway is very dangerous in times of high water. If water is flowing over the flowstone cascade at the far end of the lake the streamway will be impassable! The river passage is long and strenuous so bear the return journey in mind.

So long as the entrance lake can be crossed the fossil series will provide a memorable enough trip in its own right.

Equipment

	Depth	Rope	Fixings	Remarks
Lake	30m			Boat. 80 metre pull back line
Climb up from lake ...	4m	12m	Stal	Abseil back
Traverse	30m	50m	3 threads 5 bolts	
Casecade	12m	20m	3 bolts	
Climb out of Salle Gathier	40m	50m	6 bolts	

It is important to use STEEL karabiners for the may wire traverses.

GROTTE DE GOURNIER

Entrance porch

Fossil gallery

Lac Méduse

La Cascade Rouge

River

2nd access to river

Allée Blanche +33

Salle des Fontaines

1st access to river

+ 48 m

Terminus of fossil gallery

4th access to river

Cascades

Cascades

0   100 m

GROTTE DE GOURNIER

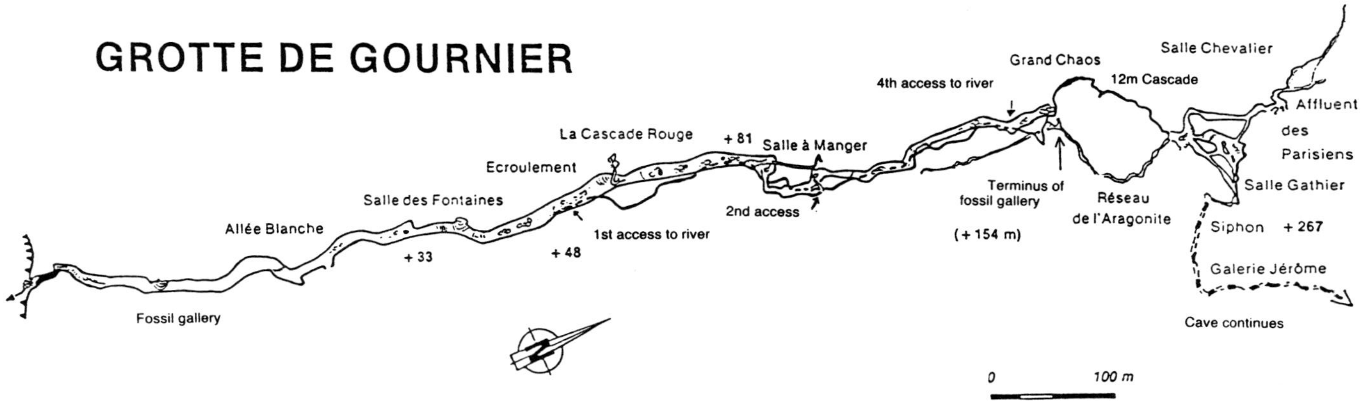

Fossil gallery

Allée Blanche

Salle des Fontaines

+ 33

Ecroulement

La Cascade Rouge   + 81   Salle à Manger

2nd access

1st access to river

+ 48

4th access to river

Grand Chaos

Salle Chevalier

12m Cascade

Affluent des Parisiens

Salle Gathier

Siphon   + 267

Galerie Jérôme

Cave continues

Terminus of fossil gallery

(+ 154 m)

Réseau de l'Aragonite

0   100 m

45

GROTTE DE BURY

Map: Carte IGN 1/25,000 No 3235 OT Autrans/Gorge de la Bourne/Parc Naturel (TOP 25). This cave is marked on the map.

Map Reference: X: 843,48   Y: 318,94   Z: 1258

Depth: −520m
Length: 4,900m

Access
From La Balme-de-Rencurel in the Bourne Gorge take the D35 road to Rencurel and carry on to the Col de Romeyere. From here turn left and follow the narrow road that climbs up the western flank of the valley to the Col du Mont Noir. Continue following this road to a junction with a road entering from the left. Continue straight on past a 90 degree bend to the left and then one to the right followed by a sweeping bend to the left. Ignore the turning to the right which comes up from Malleval and continue along, driving slowly, looking for a path that descends into the wood on the right about 250 metres further on from the turning. A cairn and spent carbide indicates the place to park. Follow the path down a shallow valley for 100 metres to the small cave entrance. The 1·5m high opening is easily found and quickly leads to a bedding chamber.

History
In 1938 the caving group of the Paris section of the French Alpine Club (C.A.F.) explored the cave to −75m stopping at some very wet squeezes. During 1950 the 'Groupe des Cyclopes' chemically enlarged the squeezes and found another passage descending to − 85m and the top of a vertical squeeze, the 'Etroiture des Cyclopes'. It was not until May 1965 that the F.L.T. after dynamiting passed this squeeze. The exploration continued to the siphon at −354m which was reached in August 1965. In June 1972 divers from the F.L.T. and the S.G.C.A.F. explored 600m of new passage to where they were stopped at −405m at a second siphon. In October 1979 F. Vergier (Darboun) and F. Poggia dived in siphon 2 and also in 3 and 4. In June 1984 J. Massehelein and S. de Vos (a Belgian diver) passed siphon 4, stopping at a cascade. In October of the same year F. Poggia and J. Massehelein were stopped at −520m by siphon 6.

Description
A short section of very shattered fossil passage leads quickly from the entrance to a narrow meandering streamway. Downstream progress is difficult due to a build up of stal. At one point a climb out of the streamway avoids a blockage and leads to P8. This is easily free-climbable as it is in a narrow rift. 200m further on the 'Puits de Confluent' (8m) is a short climb, a vertical stal flow leading to a pool. Downstream quickly leads to a siphon and the way on is to climb 7m above the siphon to a 10m flat out crawl which draughts strongly. This leads to the 'Etroiture des Cyclopes' – a 6m descending, narrow rift which at one point is only 30 cms wide! Be careful not to get hung up on SRT kit. It is recommended that your Petzl stop is put into the non-stop mode as you may find it impossible to operate the handle. Descend to the streamway beyond the siphon, which continues larger and without blockages making easy progress to deep water. After the deep water a short free climbable pitch of 3m leads to the 'Puits de l'Escarpolette'

(18m). Descend down the wall of the spacious shaft in the water. A slippery climb follows, the 'Toboggan', which leads onto the 'Puits du Piton' (10m) shortly followed by the 'Puits de Tenebres' (20m). This pitch is free hanging but in the water drops into a large boulder chamber, the 'Salle du Tenebres'.

Continue downstream to reach a junction after 150m. Follow the left hand passage. The character of the cave changes as there is now little gradient to the stream. After 400m following the stream past various obstructions reach a handline and use it to traverse out above the stream to reach a stal slope which is difficult to climb. At the top a squeeze over a boulder is followed by a descent to an 8m pitch. After the pitch an easy high level route downstream reaches a 10m pitch which rejoins the stream which is followed through pools and down cascades until it vanishes through a narrow slot. Climbing high above the stream bypasses this obstacle. Two more obstruction are passed in similar fashion. After the last barrier the roof lowers and the passage looks as though it may come to a siphon but there is a considerable distance to go to the siphon down cascades and through deep pools. Eventually the diving line of siphon 1 is met (−354m). Most cavers call it a day here but a way has been forced above the siphon and down to −405m, the end for non-divers.

Equipment

Shaft	Rope	Fixings	Remarks
P8......................	15m	2 bolts sling	Free climbable
Puits du Confluent (8m)	15m	3 bolts	
P7......................	10m	2 bolts sling	
Etroiture des Cyclopes (6m) .	10m	2 bolts	
P3......................	5m	1 bolt	Free climbable
Puits de L'Escarpolette (18m)	28m	3 bolts	
Puits du Piton (10m)	15m	3 bolts	
Puits de Tenebres (20m).....	30m	4 bolts	
P10.....................	12m	3 bolts	Avoidable by awkward climb
P8......................	12m	2 bolts	
P10.....................	13m	3 bolts	

GROTTE DE BURY

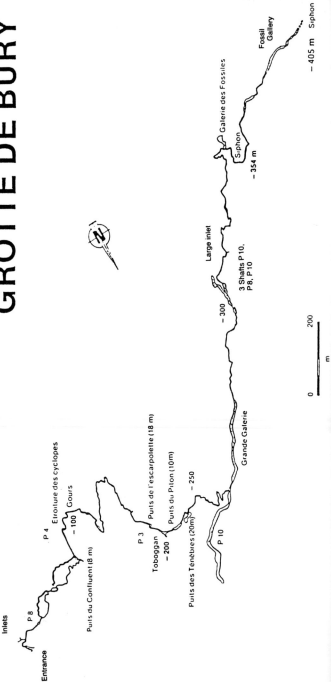

Entrance

Inlets

P 8

Puits du Confluent (8 m)

P 4

Etroiture des cyclopes

Gours

− 100

Toboggan

P 3

Puits de l'escarpolette (18 m)

− 200

Puits du Piton (10m)

Puits des Ténèbres (20m)

− 250

P 10

Grande Galerie

− 300

Large inlet

3 Shafts P 10, P 8, P 10

− 354 m

Siphon

Galerie des Fossiles

Fossil Gallery

− 405 m Siphon

N

0   200

m

GROTTE FAVOT

Map: Carte IGN 1/25,000 No 3235 OT Autrans/Gorge de la Bourne/Parc Naturel (TOP 25). This cave is marked on the map.

Map Reference: X: 848,52   Y: 323,41   Z: 880

Depth: −116m
Length: 850m

Access
Finding the path that leads to the cave is difficult! It is found by parking your car in the lay by at the end of the first new stone wall some 200 metres down from the Pont Goule Noire in the Bourne Gorge. The steep path goes up scree to the cliff face where a scramble leads up to the left to reach the large entrance. A bit of a slog!

Description
The large entrance diminishes in height and a short stoop and crawl along the mud floor emerges into the 'Grand Tunnel', a remarkable pentangular phreatic passage some 5m in diameter descending into the cave. To the left is a spectacular balcony. Descent of the 'Grand Tunnel' for 70m reaches a large chamber decorated with ancient formations. Use of a handline is recommended for the descent of the latter half of the passage. Following the passage down soon reaches the pitches. P14 and P35 descend down a huge stal to another mud floored chamber and the lowest level of the cave.

Equipment

Shaft	Rope	Fixings	Remarks
'Grand Tunnel'	55m	Boulder	Handline
P14			
P35	60m	5 bolts	

It is certainly well worth the slog up the hill to see the 'Grand Tunnel'. This is an ideal cave for anyone having a quiet day or just wanting a quick trip. It can be combined with the Goule Blanche (easily recognised by the metal walkways leading into the cave) on the opposite side of the Bourne Gorge. There is limited parking here and the cave is well worth a visit especially when it is in flood! Just beyond the road tunnel by the Pont Goule Noire a path descends to the river and the resurgence cave of the Goule Noire. Again this is most impressive in flood. A more exciting approach to the Goule Noire is to abseil off the road bridge to the river bank some 120 feet below!

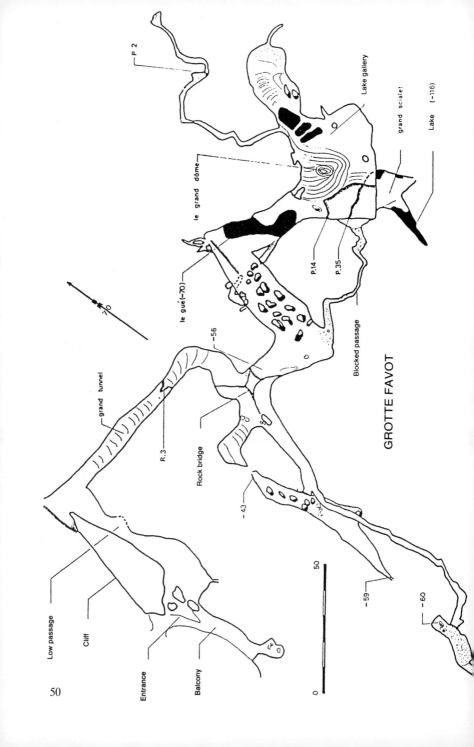

GROTTE FAVOT

50

P 2

Lake gallery

grand scialet

Lake (~116)

le grand dôme

P.14

P.35

Blocked passage

le gué (~70)

~56

grand tunnel

R.3

Rock bridge

-43

-59

-60

Low passage

Cliff

Entrance

Balcony

50

0

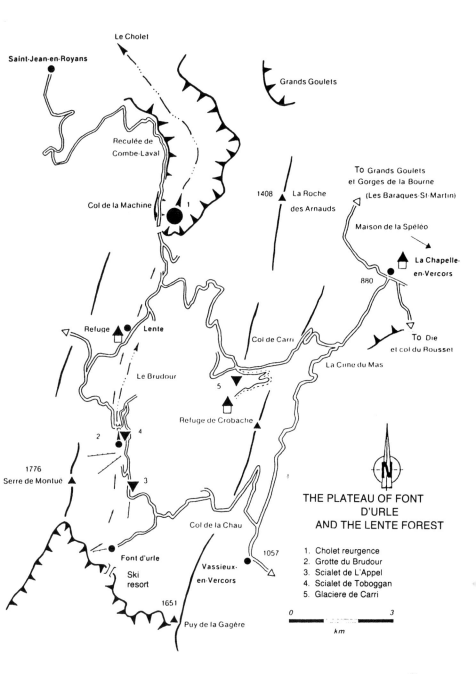

Le Cholet

Saint-Jean-en-Royans

Grands Goulets

Reculée de
Combe-Laval

Col de la Machine 1

1408 La Roche
des Arnauds

To Grands Goulets
et Gorges de la Bourne
(Les Baraques-St-Martin)

Maison de la Spéléo

La Chapelle-
en-Vercors

880

Refuge Lente

To Die
et col du Rousset

Col de Carri

Le Brudour

La Cine du Mas

5

2 4

Refuge de Crobache

1776
Serre de Montué

3

Col de la Chau

Font d'urle

1057

Ski
resort

Vassieux-
en-Vercors

N

THE PLATEAU OF FONT
D'URLE
AND THE LENTE FOREST

1. Cholet reurgence
2. Grotte du Brudour
3. Scialet de L'Appel
4. Scialet de Toboggan
5. Glaciere de Carri

1651
Puy de la Gagère

0   km   3

SCIALET DE L'APPEL and GROTTE DU BRUDOUR

Map: Carte IGN 1/25,000 No 3136 ET Combe Laval/Foret de Lente/Parc Naturel (TOP 25). The Brudour is marked on the map but the Appel is not.

Map References: Scialet de l'Appel   X: 836,02   Y: 294,87   Z: 1345 Grotte du Brudour   X: 825,62   Y: 295,90   Z: 1210

Depth: +199m
Length: 5,900m

Access
Scialet de l'Appel. Approach from Vassieux over the Col de la Chau on the D76. Continue past the turn off for Font d'Urle and continue towards Lente. 100 metres before the Carrefour du Brudour Sud take a path leading off to the right (East) to reach the cave in 100 metres. There is ample parking here off the road.

Grotte du Brudour. Approach as above but continue past the Carrefour du Brudour Sud to the Pont du Brudour some 2km past the Scialet de l'Appel parking. Park your car by the bridge and walk upstream for 10 minutes to the cave entrance.

History
The Grotte du Brudour has been known since the 13th century but the first explorations took place around 1869 with the entrance being explored by M. Gonnelle who stopped at the first stretch of deep water. 27 years later in July 1896 who else but A. Martel along with B. Delebecque followed the river for 400m. In July 1899 A. Martel returned with his friends, de Lottier and Perrin and reached the upstream siphon at +17m. It was not until 12th June 1971 that J. Dubois of the F.L.T. dived the first siphon but were stopped at a second. This was passed 3 months later by M. Chiron and R. Jean of the S.G.P.C.A.F. Later in 1971 the F.L.T. and the S.G.P.C.A.F. explored 700m of river beyond the Brudour siphon, and the start of the Reseau de Montue and the Reseau d'Urle.

The Reseau d'Urle was explored during the winter and spring of 1972 to a choke above a siphon at +92m. At the same time as the Reseau d'Urle was being explored the F.L.T., searching the surface for entrances, found the Scialet de l'Appel but were stopped at −8m by branches and loose stones. When these were cleared the F.L.T. and the S.G.P.C.A.F. were stopped by another siphon at −33m (Upstream siphon of the 'Reseau d'Urle'). M. Chiron found the junction on 13th August 1972 with the Scialet de l'Appel and the Grotte du Brudour by diving the siphon.

Following this underwater junction the F.L.T. and the G.S. Coulmes attacked the choke at -92m and on the 20th September 1972 a way through was found. This made easier access for future explorations on the siphons of the Grotte du Brudour. The 'Affluent d'Herbonnouse' (J. L. Rocourt) was climbed to +132m; the 'Affluent de Montue' 1 to +136m and the 'Reseau de Montue' 2 to +199m.

Description
Scialet de l'Appel. The 15m entrance pitch rope is belayed to the convenient tree. A small hole at the bottom of the pitch leads to a keyhole shaped passage with mud on the floor. In 30m the 'Puits de la Riviere' (13m) descends to the stream, the Urle river. Upstream sumps in 40m. It is best to traverse at high level to the third pitch (P7) at the

bottom of which is a choice of ways. Along the stream the short passage with holes in the floor leads to a deep siphon or a traverse at high level followed by a difficult 3m climb up over a rock lip (with bolt on the edge). Climb down 3m on the far side to enter a walking-size passage with large holes in the floor leads to an obvious junction (Trou Asp marked with soot on the wall).

The way on veers right to 'Puits Jocelyn' (12m). At the bottom of this follow the fossil passage ('Galerie du Shunt') to 'Puits Katia' (7m) leading back to the river. At the bottom of the pitch descend the stream to another pitch, 'Cascade Genevieve' (5m) dropping into a chaotic chamber. At a low arch join the 'Affluent d'Herbonnouse'. The route here starts off with some squeezes and quickly gains height with cascades and gours. Downstream from the junction a short fossil passage reaches the next pitch P7 where the river cascades down into the 'Salle du Carrefour'. Here, to the left, a low crawl gives access to the 'Reseau de Montue'. To the right of the chamber following the river a siphon is reached, (Siphon 1 of the Scialet de l'Appel, Siphon 2 of the Grotte du Brudour).

N.B. Descend only in settled weather. Most of the active cave would be impassable in flood and some sections would fill to the roof, notably at the junction with the 'Affluent d'Herbonnouse'.

Grotte du Brudour. The cave, protected by an iron fence, starts as a vast porch with the initial passages continuing in similar pattern, crossing various pools and lakes. As one continues upstream the passage shrinks at a canal to enlarge again at the 'Salle du Grande Chaos' where the chamber is 70m × 20m × 20m. At the far end of the chamber a fine meander carries the 'Riviere Serpentine' to the first siphon of the Grotte du Brudour.

Equipment
Scialet de l'Appel

Shaft	Rope	Fixings	Remarks
Entrance P15	20m	Tree	
		1 bolt rebelay	
Puits de la Riviere P13......	20m	4 bolts, thread	Dev. −4m
P7......................	15m	2 bolts	
Traverse and pitch	20m	3 bolts	
Puits Jocelyn P12	15m	3 bolts	
Cascade Katia P7	12m	2 bolts	Wet
Cascade Genevieve P5	10m	3 bolts	
P7......................	12m	3 bolts	

It is easy to use more rope than recommended above on the traverses so it is advisable to take 2 extra 15m ropes.

The use of an inflatable boat for the Grotte du Brudour is the most practical way of exploring the lakes and canals.

If exploring the 'Reseau de Montue' or the 'Affluent d'Herbonnouse' extra rope must be taken for the climbs.

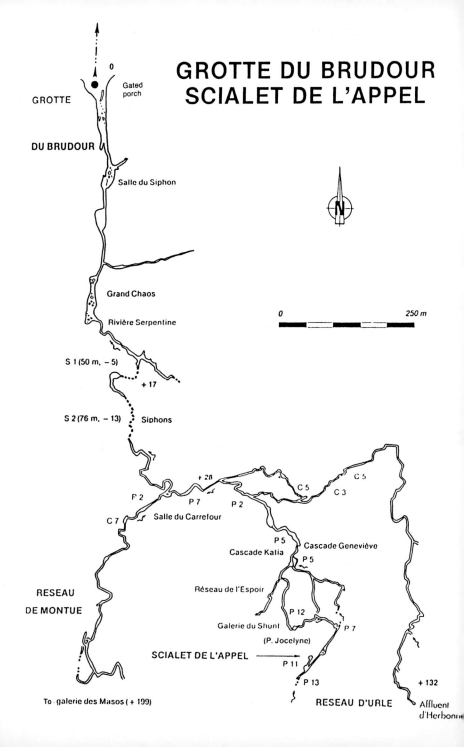

GROTTE DU BRUDOUR
SCIALET DE L'APPEL

0

Gated porch

GROTTE

DU BRUDOUR

Salle du Siphon

N

0   250 m

Grand Chaos

Rivière Serpentine

S 1 (50 m, − 5)

+ 17

S 2 (76 m, − 13)   Siphons

+ 20

P 2   C 5   C 5
                                    C 3
P 2   P 7   P 2

C 7   Salle du Carrefour

P 5
                    Cascade Geneviève
Cascade Katia
                    P 5

Réseau de l'Espoir

RESEAU

DE MONTUE

P 12

Galerie du Shunt   P 7

(P. Jocelyne)

SCIALET DE L'APPEL ⟶

P 11

P 13   + 132

To · galerie des Masos (+ 199)

RESEAU D'URLE   Affluent
                                    d'Herbonne

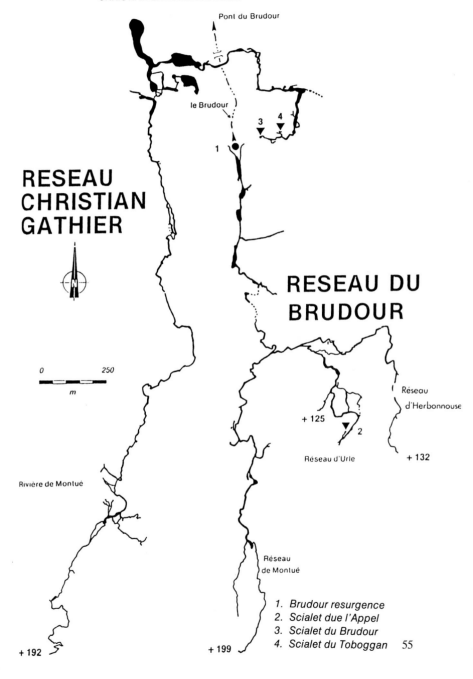

Pont du Brudour

le Brudour

3   4

1

**RESEAU
CHRISTIAN
GATHIER**

N

**RESEAU DU
BRUDOUR**

0   250

m

Réseau
d'Herbonnouse

+ 125

2

Réseau d'Urle   + 132

Rivière de Montué

Réseau
de Montué

1. Brudour resurgence
2. Scialet due l'Appel
3. Scialet du Brudour
4. Scialet du Toboggan   55

+ 192   + 199

RESEAU CHRISTIAN GATHIER

Map: Carte IGN 1/25,000 No 3136 ET Combe Laval/Foret de Lente/Parc Naturel (TOP 25).

Map References: Scialet du Toboggan   X: 835,79   Y: 295,95   Z: 1244 Scialet du Brudour   X: 835,71   Y: 295,95   Z: 1230

Depth: 309m (−107m, +202m)
Length: 9,406m

Access
From La Chapelle en Vercors take the road towards Vassieux. Shortly before reaching Vassieux the D76 turns off to the right, leading over the Col de la Chau towards the Font d'Urle ski resort. Continue down the D76 past the Carrefour du Brudour Sud. 1 kilometre further there is a small lay-by (with dumped carbide) on the left and a reasonably well defined but narrow track leading down into the forest. Follow this track until it starts to veers left and traverses to a 2 metre high cliff, at the base of which is a small (·5m x ·7m) hole emitting a strong draught. This is the Scialet du Toboggan, the normal entrance. 5 minutes from the car.

History
Martel (who got everywhere) knew of the Scialet du Brudour in 1896 and thought that it might connect with the nearby Grotte du Brudour. When the G.S.V. looked at it they too did not record any passage.

In 1974 G. Bohec and the S.C. Vizille, attracted by the current of air, were stopped at a boulder in a crawl, having followed the draught.

Members of the G.S.C. (Groupe Speleologique des Coulmes de St. Marcellin) passed the obstruction on June 19th 1975. On the 11th July the explorers came out by following the current of air via the then unknown Scialet du Toboggan, thus avoiding the famous 'Chatiere des Vizillois'.

During the following two years more than 9 kilometres were explored and surveyed. The 'Reseau de Montue' was climbed up to +202m and the siphons upstream and downstream of the river Bournette were dived.

The principal explorers were: Caillat, J. Favre-Novel, J.M. Frachet, P. Garcin, Pain, Ruel, J.P. Vincent, also G. Bohec, M. Chiron, A. Marbach and J.L. Rocourt.

Description
Although the entrance rift is free climbable, it is safer to use a 35 metre rope belayed to the tree above the hole. (There is also a thread 3 metres inside the cave entrance). Easy vertical squeezes descend to a small chamber at the bottom. Take the obvious and well worn passage down, through a number of low, muddy but short crawls, separated by larger passage, to a 5 metre climb up to a higher level where an 8 metre pitch drops down into the level of the previous passage. Beware of fixed ropes on this and the other pitches.

A short distance further on a 7 metre pitch drops into the Premier Metro, the start of the large passages. Follow these through marvellous formations until a 10 metre pitch drops into the stream. This is followed downstream until just before a sump where by climbing up through boulders you emerge into the very impressive and large Salle des Tenebres. N.B. Take note of the exit from this chamber as it can prove difficult to find on the return journey.

At the bottom of the Salle des Tenebres a low passage briefly joins the streamway before ascending into the bottom of the Galerie Geante which climbs up to just 35 metres below entrance level! Keeping to the left here, instead of following the Galerie Geante, a draughting squeeze leads to a 10 metre drop (where you need to fix a handline tied around boulder on the lip). Below this drop there is a series of large passages. The one with a large waterfall cascading into it is the start of the Riviere de Montue which requires scaling gear as the climbs are now derigged.

Equipment

Shaft	Rope	Fixings	Remarks
Entance	35m	Tree Thread 3 bolts	
5m climb	10m	Stal	Descend far hole
P8	10m	Spike 1 bolt	
P7	12m	3 bolts	
P10	12m	3 bolts	Access to streamway
P7	10m	2 bolts	Access to Salle de la Cascade

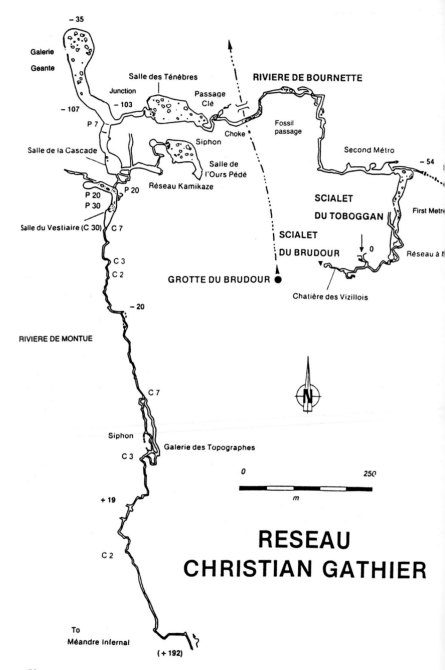

Galerie Geante

− 35

Salle des Ténèbres

Junction

− 103

− 107

P 7

Passage Clé

RIVIERE DE BOURNETTE

Choke

Siphon

Fossil passage

Second Métro

− 54

Salle de la Cascade

Salle de l'Ours Pédé

Réseau Kamikaze

SCIALET DU TOBOGGAN

First Metro

P 20

P 20

P 30

Salle du Vestiaire (C 30)

C 7

SCIALET DU BRUDOUR

Réseau à I

C 3

C 2

GROTTE DU BRUDOUR ●

Chatière des Vizillois

− 20

RIVIERE DE MONTUE

N

C 7

Siphon

C 3

Galerie des Topographes

0   250

m

+ 19

RESEAU CHRISTIAN GATHIER

C 2

To Méandre Infernal

(+ 192)

58

GLACIERE DE CARRI

Map: Carte IGN 1/25,000 No 3136 ET Combe Laval/Foret de Lente/Parc Naturel (TOP 25).

Map Reference: X: 838,80   Y: 297,84   Z: 1290

Depth: −193m
Length: 380m

Access
From La Chapelle en Vercors take the road towards Vassieux and turn right at La Cime du Mas on the D199 to the Col de Carri. Leave the car at the summit and follow the track SSE towards the Crobache Hut. After 150m by the first left hand bend climb up to the right (S) on a well marked path. Pass the Scialet de Carri shortly after but continue uphill passing a 10m shaft until you reach, 150 metres further on the Glaciere, a large opening with a ledge on the right. (20 minutes from the car).

History
The cave has always been known but the passages at the foot of the entrance shaft were not explored until September 10th 1959 by J.C. Soletty (S.C.V.). A few descents later and −64m was reached but on October 3rd 1959 Aldo Sillanoli of the S.G.C.A.F. reached −109m.
In 1969 (J. and M. Lamberton, J. Bonnet, B. Begou, D. Glauque, P. Garcin, J. P. Moratal) members of the G.S.V., starting using dynamite and found the Reseau de Valentinois. They reached −124m in May/June 1970.
The cave was visited again in 1974 by the F.L.T. (J. Dubois, P. Pesquet, J.L. Rocourt) dynamiting a window at −102 . Tight descending squeezes led to P43. J.L. Rocourt descended on April 27th 1974 but was stopped at −173m. In 1979 the 'Speleos Grenoblois' of the C.A.F. returned to the cave and on November 25th G. Kirkor, F. Leclerq and B. Lismonde removed an obstruction at the bottom of P43 to squeeze through and reach the bottom at −193m.

Description
The large daylight shaft is 15m deep and has neve at the base. A low hole connects with a chimney. A window at head height leads through onto the head of a pitch, P10. In mid-descent another window gives access to a climb up to the head of a series of shafts, P10, P5, P15, P9, P5, P11, P33, P14. The bottom of P14 was formerly the end of the system. A pendule at −3m on P14 gives access to a series of tight vertical squeezes to a fine P43. For the brave and devoted caver the bottom of the cave is reached by following a current of air in fairly complex and tight passages.
The trip to the end of P14 makes a really fine introduction to SRT.

Equipment

Shaft	Rope	Fixings
P15...........................	30m	4 bolts
P10...........................		
P5...........................		
P15...........................		
P9...........................		
P5...........................		
P11...........................		
P33...........................		
P14...........................	135m	18 bolts
R4	10m	2 bolts
Block	10m	2 bolts
P43...........................	60m	4 bolts
P8...........................	12m	3 bolts

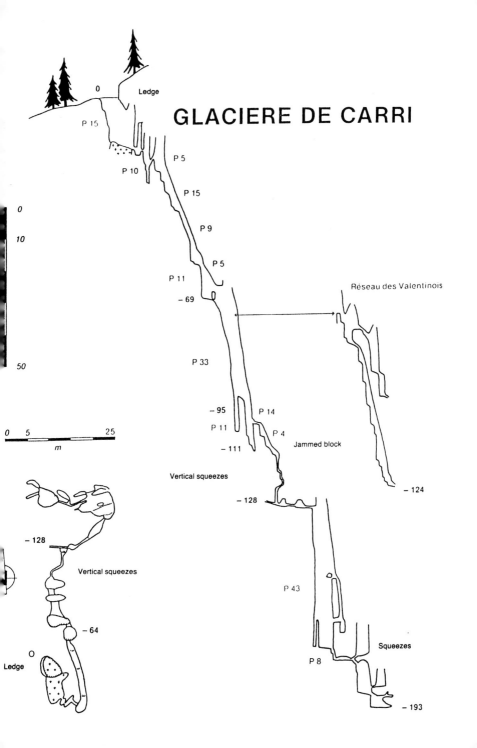

GLACIERE DE CARRI

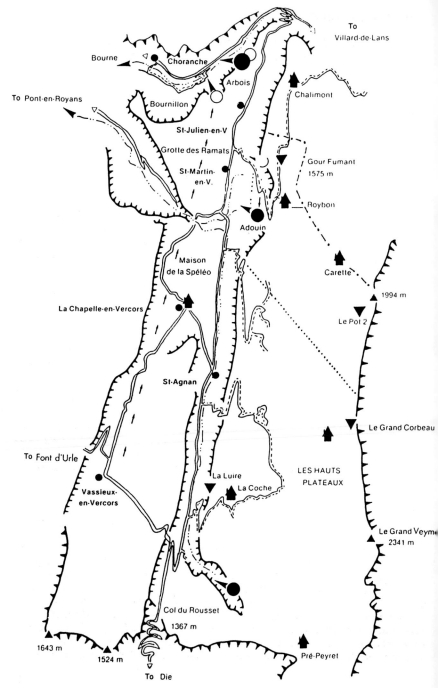

THE ARBOIS – BOURNILLON – LUIRE HYDROLOGICAL
SYSTEM AND THE L'ADOUIN BASIN.

0   4 km

GROTTE DE LA LUIRE

Map: Carte IGN 1/25,000 No 3236 OT Villard de Lans/Mont Aiguille/Parc Naturel (TOP 25). The cave is marked on the map.

Map Reference: X: 844,40   Y: 292,53   Z: 875

Depth: 513m (−450m, +63m)
Length: 17,799m (upstream 9,713m; downstream 8,086m)

Access
From St-Agnan-en-Vercors take the road, D518, towards the Col du Rousset. Follow signs to the show cave and take the D622 that leads to the car park. As the cave is partly a show cave it is necessary to ask permission which is freely granted to descend the cave. However, it is very important to arrange your trip the day prior to your visit. If there is any doubt at all with the weather you will not be allowed to descend. Generally this will be known when you arrange your trip the day before but if there is any doubt no one will be allowed to enter. The whole known cave fills to the roof very quickly, especially in times of snow melt when it is not unknown for 40 cubic metres of water a second to resurge from the cave entrance! (Look at the pictures by the pay booth.)

The system is epi-phreatic, meaning that it fills up from below and in normal conditions the water found in the system drains out through the Grotte de Bournillon. The walk to the show cave entrance is 3 minutes and you need to arrive by 10.00 a.m.

History
The first explorations were made in 1896 by the local people of La Chapelle-en-Vercors when they descended to −86m in the Puits Bis. In 1898 O. Decombaz was stopped at −90m at the top of the 'Grand Scialet'.

The next explorers, the S.G.C.A.F., started their explorations in 1936 where they descended the 'Grand Scialet' to −186m and in 1945 reached −218m. During World War 2 the cave entrance was used as a hospital by the resistance.

In the 1950's the G.S.Valentinois explored the rest of the system. In 1952 they explored the passages that bear the name and year, 'l'Aval 52', in 1953 the 'Galerie de Noel' and in 1959 the 'Galerie des Nenuphars'. During 1961-62 the active upstream section was explored. After a 15 year break the G.S.V. reached the terminal siphon the 'Crepuscule des Dieux' (Twilight of the Gods) at −450m.

Description
The normal route of access to the lower cave is by the 'Grand Scialet'. Although there is an alternative – the Puits Bis – it is not described here.

Climbing over the railings of the show cave one descends a large shaft (15m) rigged with a fixed ladder and rope. It leads to a 20m scree slope. Take care not to damage the electric cables here. Fixed iron ladders descend to the lower cave. Despite new sections of ladder they do not inspire confidence so it is politic to use the fixed ropes, or, using the fixed ropes as a guide, fix your own, abseiling in and climbing out using a self-lining jammer. Ensure that you take a FULL SRT kit with you – just in case! The shaft is fault aligned and stonefall can be a hazard. At the base of the shaft a pebble and boulder strewn floor has obvious erosion features due to the violent up-welling of water. A loose descent beneath the boulder floor leads into a low and bouldery passage to the right.

'Pseudo-siphon 2', a duck in normal conditions, is soon met. Beyond, the passage quickly increases in height. On the right another duck, 'Pseudo-siphon 1', leads to a rising passage in a superb circular phreatic tube to enter a sandy chamber with 3 ways on. Going right leads back to the approach passage, straight on soon ends, whilst to the left the passage continues upwards along the 'Galerie du Courant d'Air'.

Between the two ducks a bouldery passage leads immediately to the main junction with large passage heading both ways. To the right the passage slopes gently down to a traverse round a hole in the floor after 300m or so. Shortly after this the passage divides, the right hand branch slanting up to a shaft, P15, on to P10 and then the 25m 'Puits Terminal'. Continuing downwards to the left at the main junction, a large passage enters the 'Galerie des Gours'. After 900m the passage ends at an awkward climb down into 'Les Salles' which descends steeply to a junction. Left, a small passage goes to −311m, right goes to −350m the lowest attainable point, a thought provoking place especially if water starts to well up! It rises at 1 metre per minute!!

Equipment

Shaft	Rope	Fixings	Remarks
P12	15m	3 bolts	fixed iron ladder fixed rope
P22	30m	2 bolts	fixed iron ladder fixed rope
'Grand Scialet'	120m	8 bolts	fixed iron ladders fixed rope

GROTTE DE LA LUIRE

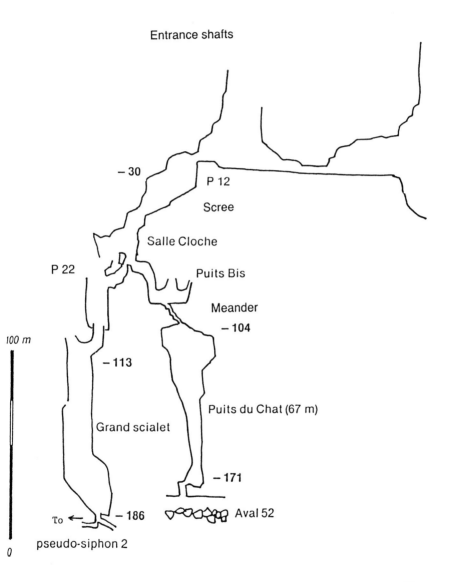

Entrance shafts

− 30

P 12

Scree

Salle Cloche

P 22

Puits Bis

Meander

− 104

100 m

− 113

Puits du Chat (67 m)

Grand scialet

− 171

To ← − 186   Aval 52

pseudo-siphon 2

0

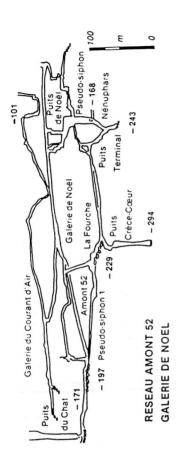

GROTTE DE LA LUIRE

RESEAU AMONT 52
GALERIE DE NOEL

RESEAU AMONT 52
GALERIE DE NOEL

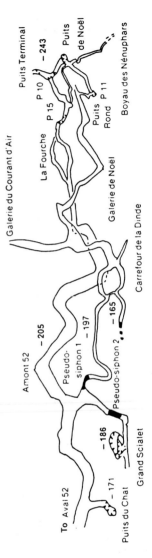

66

GROTTE DE BOURNILLON

Map: Carte IGN 1/25,000 No 3236 OT Villard de Lans/Mont Aiguille/Parc Naturel (TOP 25). The cave is marked on the map.

Map Reference: X: 843,79   Y: 310,61   Z: 418

Depth: −65m
Length: 5,950m

Access
Some 600 metres beyond the turning to the Choranche Show Caves, off the road running through the Bourne Gorge, a sharp turning to the left leads to the Hydro Electric power station. Either park the car before the bridge over the river or, if there is space, 100 yards beyond the power station on some flat ground close to a huge water pipe. It is possible to approach the power station from Choranche itself by following the minor road that winds its way through Vezor to the same parking place before the power station. A well defined path climbs up through thick vegetation, soon passing under the huge pipe which carries water to the power station across some scree, clearly visable from the main road, to reach the massive entrance porch after 25 minutes. An impressive spot.

History
In September 1897 Decombaz, Millon, Pellerin and Chastel explored the cave as far as the 'Village Negre'. One month later, Decombaz accompanied by Perrenot, Argencon and Chabert continued beyond the 'Village Negre' to the 'l'Aiguille de Metro' and were stopped by a siphon. In September 1942, Dussere continued beyond this but soon was stopped by water. A month later, Bourgin and Penelon discovered the 'Galerie Laterale' but stopped at the siphon mentioned above. They also discovered that the 'Galerie Superiere' linked with the main porch at the 'Village Negre'. A little later Pommier found the access to the passage that bears his name. In September 1955, M. Letrone helped by the Clan des Tritons passed beyond the siphon to the upstream of the 'Galerie Laterale' but stopped at some pools. Seven years later G. Michel found the siphon dry and found the start of the labyrinth. In 1971 the conditions allowed the G.S. des Coulmes and the S.G.P.C.A.F. to explore the labyrinth. In the same month R. Jean and M. Chiron carried on for 80m in the left branch of the 'l'Aiguille de Metro' siphon. In April 1973 B. Leger and J.L. Camus dived the siphon for 225m to a depth of −34m. 10 years later F. Poggia went 200m further whilst another year later B. Leger found a tight meander 15m further on. The right hand branch of the siphon was dived during the same period by R. Jean, D. Andres, C. Touloumdjian and F. Poggia.

During the dry Autumn of 1985 M. and F. Chiron returned to the Labyrinth and found at the bottom, on the 20th October, a passage giving access to the 'Salle des Centaures' which continued to two more sumps, Alpha to the right and Beta to the left at +65m from the entrance. In November 1985 F. Poggia dived Beta siphon but was stopped by yet another siphon. J.L. Camus dived for 260m in siphon Alpha.

Description
If this is your first foriegn cave it is a good cave to start on. The large passages and boulder hopping will get you in the frame of mind for other caves in the area.

The 80m high entrance porch is the largest in France! The entrance porch sometimes holds a lake but the way is obvious along the high level path. Cross over the footbridge at the far end of the porch and go to an eyehole through which the stream – or river, depending on the weather – flows. If there is no water flowing then this is the best way of entering the cave but it can easily and VERY quickly become impassable. It is prudent therefore to acquaint oneself with the alternatives. These are not easy to find from within the cave without prior knowledge.

These alternatives are high up on the boulder slope to the right of the entrance porch and are obvious from the outside. The first one leads via a high level series and a squeeze to enter the main passage by the 'Village Negre', whilst the other is up the boulder slope just before the footbridge and leads onto a ledge high on the wall of the main passage. A traverse and short climb down (rope for novices) leads to the boulder floor of the main passage just inside the eyehole.

From the eyehole ascend steeply over boulders. It is possible to hear the stream far below. Route finding to reach the end of the cave is no problem through the beautifully sculptured phreatic passages. However, two points need to be taken into account. Firstly, having climbed up all the boulders you come to an area where the roof lowers and the stream flows over the boulders on the floor. No problem in normal conditions but this section can become impassable in high water conditions and can sometimes sump. Secondly, after a long period of dry, settled weather (3 – 4 months), the 85 series of low passages at the far end of the cave are occasionally open and there is a very great risk of getting caught out, by a very small rise in water levels. Take great care if you do decide to go to the end of the cave and pass beyond the level of sand deposits that mark the level of the terminal siphon in normal conditions; getting caught out below WILL prove to be FATAL. Beyond these low passages and 30 metres before the end of the Labyrinth a window at the base of the east wall gives access to a very rough crawl, 'Chiron's Way', which opens out to the large 'Salle des Centaures' (30m × 10m). More crawls are entered after you have climbed to the top of the very sloping chamber to enter a chaotic chamber known as 'Minos Center' at the start of a beautiful passage – 'les Champs Elyseens' (10m high and between 4m and 8m high). This passage is finely sculptured. Above a second step the passage, 'Table du Pluton' +65m splits. Right goes to the 'Alpha' siphon and the left goes along a 6m diameter passage to 'Beta' siphon.

N.B. In times of flood 80 cubic metres of water a second can surge from the cave!!

GROTTE DE BOURNILLON

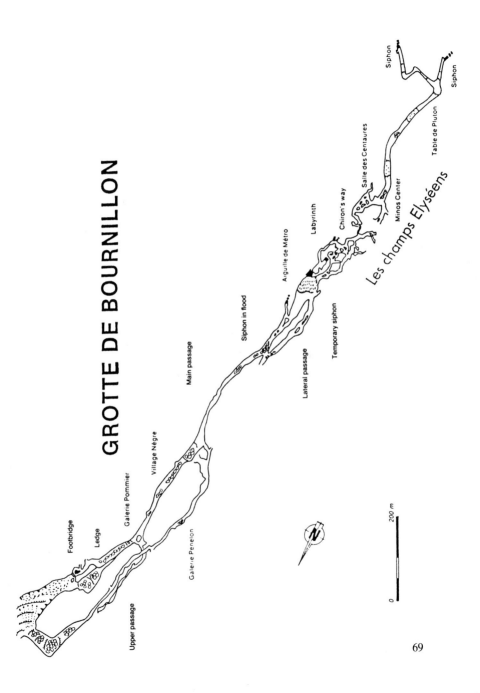

Footbridge

Ledge

Galerie Pommier

Village Nègre

Upper passage

Galerie Penelon

Main passage

Siphon in flood

Aiguille de Métro

Labyrinth

Lateral passage

Temporary siphon

Chiron's way

Salle des Centaures

Minos Center

Les champs Elyséens

Table de Pluton

Siphon

Siphon

N

0   200 m

GOUR FUMANT

Map: Carte IGN 1/25,000 No 3236 OT Villard de Lans/Mont Aiguille/Parc Naturel (TOP 25). The cave is marked on the map.

Map References: Gour Fumant X: 847,15 Y: 307,45 Z: 1270 False Gour X: 847,15 Y: 307,48 Z: 1270

Depth: −163m
Length: 2,203m

Access
From Villard de Lans take the D215c as if going to the Scialet du Trisou until the large grassy depression of the Herbouilly pasture is reached. At the southern end a track leads off left into the pasture where cars can be parked. If the track is dry it is possible to park almost at the cave entrance! Coming from Saint Julien-en-Vercors take the road as for the Scialet du Trisou until the southern end of the Herbouilly pasture is reached and then continue as above.

History
The cave was first entered in August 1936 by A. Bourgin, P. Chevalier and F. Phafl. They went as far as −56m and then onto −95m. In October 1936 the river was reached and followed to − 120m until a lake stopped further progress.

In May 1953 the Clan des Tritons (Lyon) found the Reseau du Dragon Chinois. In September 1961 the S.S.R. (Romans) passed the lake at −120m but were stopped some 10m further on by a tight hole. It was 10 years before this tight squeeze was passed to give access to the Reseau 1971 which descends to −163m. (M. Chiron, S.G.A.C.A.F).

Description
The best entrance is the False Gour which is 30 metres north of and 10 metres lower than the other entrance. A slippery boulder slope leads immediately to a 17 metre pitch from the bottom of which an easy passage leads to the second pitch of 9 metres. At the bottom of this pitch is the 'Big Gallery' with a choice of routes.

The usual way is straight on with your back to the wall at the bottom of the pitch to a 15 metre pitch. A further pitch of 9 metres leads to a junction with the second way from the 'Big Gallery' (which is to the left and through a 'letter box' and on to a constricted pitch of 11 metres). Traversing out for 12m a pitch of 12 metres is descended. An awkward 6m climb (handline for novices) leads to another pitch of 9 metres in a rift giving access to the lower cave with its active streamway. 200 metres beyond the pitches, the 'Reseau du Dragon Chinois' is entered by a climb up on the left to low beddings and larger fossil cavities containing formations. Beyond the climb, the main stream backs up behind a stal dam. The stream in high water backs up a long way and progress can only be made along the roof! In low water the stream disappears under the stal dam and for most of the distance knee deep wading suffices although you will need to swim the deep pool at the end to the blockage at the stal dam. Here a tight wet squeeze leads into the Reseau 1971. Up to the right stal flows are ascended to −76m but down to the left bridging and chimneying reaches the end at −163m.

N.B. It is inadvisable to attempt the lower wet passages in wet or unsettled weather.

The descent via the Gour Fumant starts off as a scramble down into a bouldery chamber followed by climbs down of 6m and 4m and then by pitches of 9m, 6m and 11m to reach the 'Big Gallery'.

Beyond the entrance to the Reseau du Dragon Chinois the cave is very different from the entrance series, being tight in places, wet and muddy. For this reason it is not often bothered with.

Equipment
False Gour

Shaft	Rope	Fixings
P17	20m	3 bolts
P9	15m	3 bolts
P15	20m	3 bolts
P9	15m	2 bolts
Traverse	15m	3 bolts
P12	15m	3 bolts
Handline	10m	2 bolts
P9	15m	2 bolts

Gour Fumant

Shaft	Rope	Fixings
Climb	10m	2 bolts
Climb	10m	2 bolts
P9	15m	2 bolts
P6	10m	2 bolts
P11	15m	2 bolts

If tackling the lower cave take a 10m handline.

GOUR FUMANT

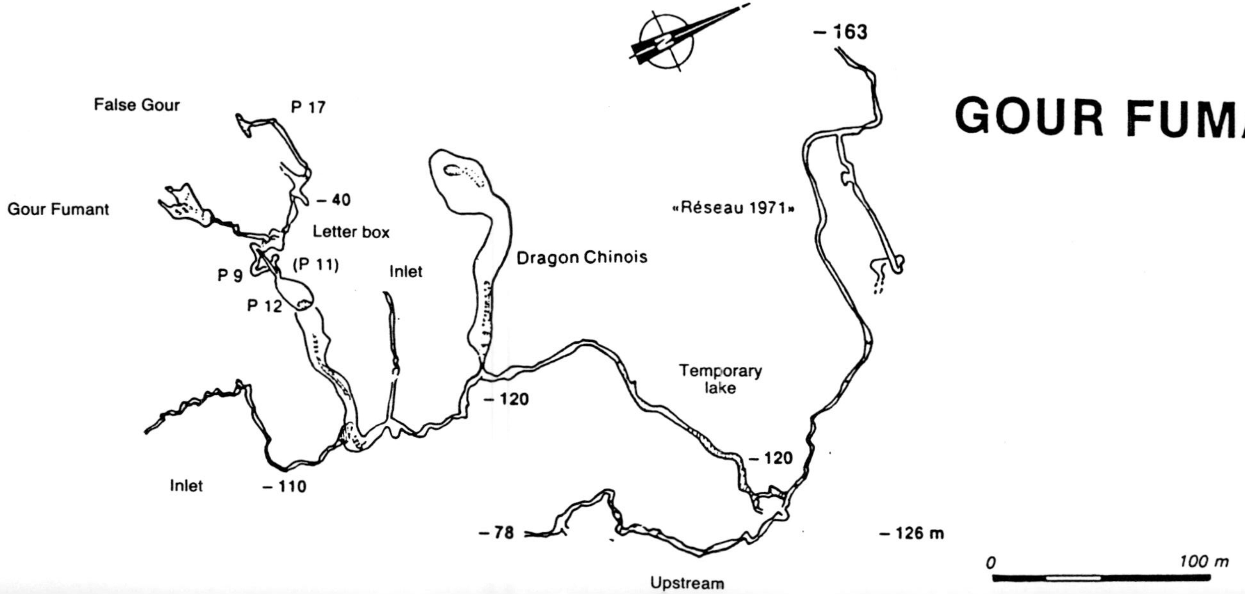

- 163

"Réseau 1971"

Dragon Chinois

Temporary lake

- 120

P 17

- 40

Letter box

Inlet

(P 11)

P 9

P 12

- 120

False Gour

Gour Fumant

Inlet

- 110

- 126 m

- 120

- 78

Upstream

0   100 m

GOUR FUMANT

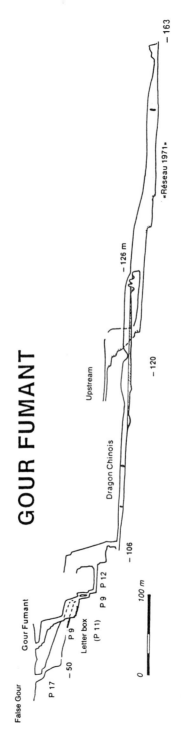

False Gour

Gour Fumant

P 17

— 50

P 9

Letter box
(P 11)

P 9   P 12

Upstream

Dragon Chinois

— 106

— 120

— 126 m

«Réseau 1971»

— 163

0   100 m

POT DU LOUP

Map: Carte IGN 1/25,000 No 3236 OT Villard de Lans/Mont Aiguille/Parc Naturel (TOP 25). The cave is marked on the map.

Map Reference: X: 847,33   Y: 307,03   Z: 1287

Depth: −94m

Access
Approach as for the Gour Fumant but continue to the southern edge of the Herbouilly Pasture and a picnic area. From here a dead-end track leads directly to the hole. Alternatively, from St. Julien-en-Vercors take the road as for the Gour Fumant until the picnic site sign is seen on the right. Take this track to the site and the dead end track that leads directly to the hole.

History
The hole was discovered by A. Bourgin in 1936.

Description
The large entrance shaft of 19m is well bolted; there are some 23 sleeves to choose from! At the bottom of this pitch a 3m climb (2 bolts at top) enters a 20m passage that leads to a series of pitches, 37m, 12m, 2m, 10m and 12m, to reach the final chamber.

Equipment

Shaft	Rope	Fixings
P19 Entrance	25m	Tree, 3 bolts
P3	6m	2 bolts
P37, P12, P2, P10 and P12	120m	20 bolts!

GROTTE DES RAMATS

Map: Carte IGN 1/25,000 No 3236 OT Villard de Lans/Mont Aiguille/Parc Naturel (TOP 25). The cave is marked on the map.

Map Reference: X: 846,17   Y: 307,46   Z: 1030

Depth: 86m (+59m, −27m)
Length: 2,500m

Access
From Saint Julien-en-Vercors take the road towards Saint Martin-en-Vercors. Midway between the two villages a turn off to the left at La Gratte takes the forest road, D221, heading for the Herbouilly pasture in the direction of Villard de Lans. Some 1,800 metres from the turning the road crosses over a small ravine. Park the car here and walk steeply up for 150 metres following the stream bed, which is normally dry, to where a small porch is seen to the right of the stream bed.

History
In 1890 E. Meller found the cave and entered as far as the first duck. In 1902 O. Decombaz continued the exploration but halted at the start of the next duck. In September 1937 J. Vignon passed this second duck and found a larger passage. In October of the same year he explored 450m of passages with A. Bourgin and the S.G.C.A.F. de Paris. In February 1938 the upstream siphon was reached (+8m and 900m from the entrance), by P. Chevalier and Chaderson (G.S.A.P.). In 1953 the Clans des Tritons (Lyon) passed the drained siphon and found a climb of 11m and explored another 600m of new passage.

Description
The low entrance drops down to a joint. Turn right. Beyond a duck (often dry) follow 500m of rift passage, which is heavily corroded, to a small stream which disappears into a siphon 720m from the entrance (+8m). A serious climb of 11m up the stal, with a squeeze at the top, bypasses the siphon. Rejoin the stream to the 'Salle de Noeud', where a short climb leads up to a junction. (To the right the passage sumps). Go left to reach a chamber and enter a very shattered section which leads into the Bourgin Gallery (970m from the entrance, +40m). A passage to the left goes along clay fill for 200m before climbing up, whilst a passage to the right leads to where 3 passages meet in a chamber after avoiding a 15m pitch. The front one gives access to the lower passages and the 'Galerie de la Verna'. To the right is the 'Galerie des Excentiques'.

WARNING: The 11m climb is quite hard but has good protection for the hard bit – a large stal thread – (if a rope and large sling is taken), especially for non climbers. A 25m rope is recommended so that it is easy to abseil back. A fall from the top, where the squeeze is, would be very serious. There is a good belay in a tiny chamber beyond the squeeze. A visit to this cave requires dry and settled weather as it fills up very alarmingly. Even a small rise in water levels would be sufficient to seal parts of the cave!!

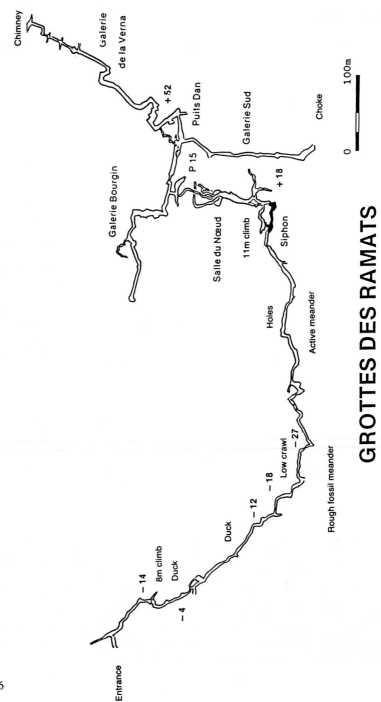

GROTTES DES RAMATS

Chimney

Galerie de la Verna

+ 52

Puits Dan

Galerie Sud

Choke

P 15

+ 18

Galerie Bourgin

Salle du Nœud

11m climb

Siphon

Holes

Active meander

Low crawl

− 27

− 18

Rough fossil meander

− 12

Duck

− 14

8m climb

Duck

− 4

Entrance

100m

0

76

SCIALET DU POT 2

Map: Carte IGN 1/25,000 No 3236 OT Villard de Lans/Mont Aiguille/Parc Naturel (TOP 25).

Map Reference: X: 851,03   Y: 300,73   Z: 1740

Depth: −319m

Access
The major difficulty of doing Pot 2 is finding it! The entrance is notoriously difficult to find and careful use of the map is essential. An altimeter is also very useful, especially if the cloud drops.

The shortest approach is via Darbounouse, the Pre de Rey Blanc and into La Purgatoire. (This last bit is very aptly named!) Up and down small cliffs and stepping over huge clints and grikes, following at first orange markers then red ones, takes you to a small ridge overlooking a huge depression to the right. Pas Morta is seen ahead and is recognisable by a large white cliff facing E. In the depression are the remains of a wooden hut. Pot 2 is found some 300 metres W of this hut beyond the rim of the depression and close to ULSA '82 signs (Scialet ULSA 1). The hole is ringed with fir trees, along with a million others in the area and the remains of a wire fence! (3·5 hours.)

Other ways of reaching Pot 2 are:- (1) via the Pas Morta which is reached from St. Andeol. From the pass (1889m) take a compass bearing of 331 degrees and follow this for 700 metres. Apparently this is the surest way of finding the hole but it is also the most strenuous. (2) From Correncon, via the Carette and the ruin of the Cabane du Grand Pot. From the ruin go S for 1·5 kilometres passing the Pot du Rey Blanc en route to join up with the route from Darbounouse.

History
Pot 2 was discovered on the 10th July 1968 by a very young team from the A.S.V. de Villard de Lans whilst prospecting for new holes in the lapiaz of the Purgatoire. That day the shaft was descended to −150 metres, by ladder!! Firstly Regis Piccavet, aged 14, went to −30 metres, then Daniel Bertrand to −150 metres.

On the 12 July J.M. Burlet, aged 15, descended by winch to −270 metres. The following day he descended to the bottom of the shaft, again by winch. The shaft is the largest in the region and one of the deepest in the world.

Description
Fixing the rope to the convenient tree on the rim of the small dimensioned (2 – 3 metres wide) shaft. The first pitch is 12 metres, followed by one of 4 metres to a small ledge overlooking the remaining drop of 303 metres. The next 20 metres remain narrow. The main shaft opens out considerably after passing a window at −40 metres. It is necessary to take care in the rigging as the existing bolt anchors are camouflaged in the fossil encrusted wall making them difficult to find, but eventually a comfortable niche is reached at −214 metres. Between −140 metres and −190 metres the shaft reaches its widest point (8 metres by 16 metres). A small hole in the floor of this niche gives access to a shaft exceeding 40 metres in depth. Stones thrown down it do not re-appear in the main shaft, but can be seen passing a window 30 metres lower. Continue down the final

100 metres of the main shaft where a bridge divides the shaft for 20 metres at −245m. Do not go down the left as there are no hanger placements but descend to the right. The base of the main shaft is 4 metres by 14 metres.

Allow 5 – 6 hours for a party of 3. There is no objective danger and no problem with water. There is no surface water so take some with you, some to drink and some for your carbide.

Equipment There are at least 3 gear lists but the one given below offers a rough guide

Equipment
There are at least 3 gear lists but the one given below offers a rough guide.

Shaft	Rope	Fixings
P12	16m	Tree, 1 bolt
P4/P303	340m	17 bolts
		Re-belays at −18, −74, −100, −170, −197, −214, −244 and −298

Other re-belays exist but need careful searching for!

SCIALET DU POT 2

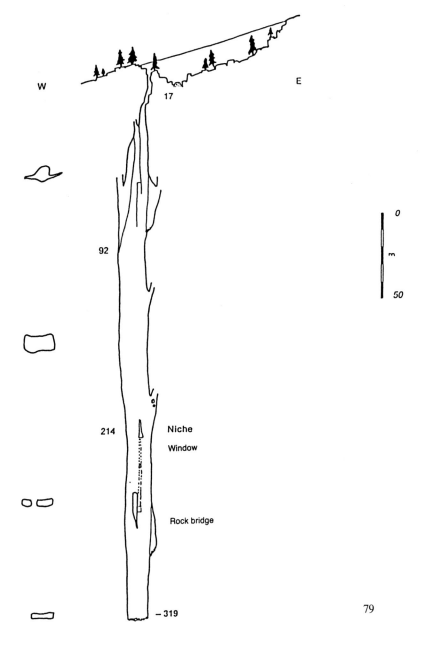

W   E

17

92

214   Niche

Window

Rock bridge

0

m

50

− 319

79

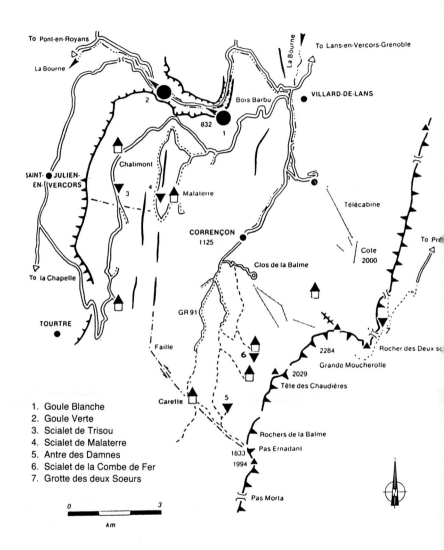

1. Goule Blanche
2. Goule Verte
3. Scialet de Trisou
4. Scialet de Malaterre
5. Antre des Damnes
6. Scialet de la Combe de Fer
7. Grotte des deux Soeurs

ANTRE DES DAMNES

Map: Carte IGN 1/25,000 No 3236 OT Villard de Lans/Mont Aiguille/Parc Naturel (TOP 25).

Map Reference: X: 851,10 Y: 303,00 Z: 1670

Depth: −723m
Length: 2,500m

Access

From Correncon follow the GR91 to the Chalet de Carette. From here head SE towards the ruined Cabane du Grand Pot. Some 200m before the Cabane turn left on a path that leads to the Glaciere du Clos de la Fure (Glaciere marked on the map). Arriving in the fault by the Glaciere go right, NNE, along a line of small cairns for 150m to find the hole on the left at the foot of a 5m high escarpment. A current of air issuing from the small entrance confirms it is the right place.

History

The cave was discovered in July 1982 during a prospecting trip by the S.C. Fontanil. After many visits the entrance was finally opened on the 17th September. By December C. Pomot and G. Sibue reached −570m. The lowest point was reached in 1983 when F. Poggia dived the two terminal siphons which ended almost immediately at some squeezes at −723m.

Cavers from Aix en Provence and the S.C. Fontanil followed the upstream passage to −490m. Since then cavers from the Drome region and the S.C.V. continued explorations to reach the end at −220m.

Description

The shored up entrance gives access to a meander ·5m − 1m wide and 5m high. The first large pitch (P60) which follows a 5m climb, is broken by a ledge at 9m. At the bottom of the pitch a larger meander is covered in moonmilk. After two 5m climbs a 30m pitch drops onto P205, 'le Goudrix'. This 15m diameter shaft descends to a final superb drop of 60m. (The shaft is impassable at times of flooding). After 12 m a pendule gives access, via a window on the right, into the dry part of the shaft for the next 100 metres. At the base of the shaft a climb of 5m reaches the top of P80, 'l'Indomptable'. This is similarly very dangerous in times of flooding! A short (4m) traverse allows one to descend the far side of the shaft away from the water.

There is a drier way of descending this pitch. However this is ONLY if you are going to explore the upstream passages. Descend 12m then pendule into a smaller but pretty series of shafts ('Puits du Colloque', P12, P12, P8, P34 and P8). P34 emerges into a short and horrible crawl before reaching the final P8. By going right at the bottom of this pitch and climbing up a small stone shoot you enter a meander ('Galerie du Bivouac'), which after a 100m enters a large (40m × 50m) chamber at −490m. Into this chamber a beautiful cascade enters from the upstream passages which are reached by a 20m climb up (rope in place − but take one just in case!) to enter a pretty fossil meander. The stream is regained some 150m further on. Follow meanders via numerous short climbs to reach −220m. Go left at the bottom of the final P8 to reach the base of P20 coming from 'l'Indomptable'.

At the bottom of 'l'Indomptable' a short passage gives access to the head of P20. Descend 6m and pendule to reach and go down P13. The descent is followed by a climb of 4m into a meander which leads to a chamber of boulders. A high window (4m up then 4m down) gives access to the 'Galerie Titan', some 8m to 10m wide and 4m high. After 100m a climb of 5m enters a huge chamber (60m × 20m) which on the left hand side has a stream entering through boulders.

The 'Galerie Titan' has some fine formations and ends at a whistling hole at −570m in a huge boulder choke. 30m before this, on the right hand side of the passage, there is a smaller passage followed by two squeezes (improved!) which give access to a meander with fine formations. Alternating meanders and stony passages are descended to reach −720m.

Equipment
To reach −720m

Shaft	Rope	Fixings	Remarks
P60	65m	Piton, 5 bolts	P9 + P51
			Deviation −30m
P5	8m	2 bolts	
P5	8m	2 bolts	
P30	35m	4 bolts	Deviation −12m
P205	230m	Natural, 2 bolts	Head of pitch
		1 bolt	−12 pendule to dry part of shaft
		1 bolt	−13
		1 bolt	−27
		1 bolt	−53
			Deviation −60
		1 bolt	−70
		1 bolt	−81
			Deviation −95
		1 bolt	−105
		1 bolt	−114
		1 bolt	−137
		1 bolt	−147
P5	10m	2 bolts	
P80	100m	6 bolts	Traverse to pitch head
		1 bolt	−6 pendule to opposite wall. Bolt under the window
		1 bolt	−22
		1 bolt	−48
		1 bolt	−57
Traverse to P20		Natural	
and down		2 bolts	Pendule at −6 to reach P13
P13	30m	1 bolt	
P4	7m	Natural	In place?
P4	10m	Natural	
P5	10m	1 bolt	Devision −4
P6	10m	1 bolt	
P5	10m	Natural	Do not use bolt sleeve on left!
P3	5m	Natural	For return!
		1 bolt	
P5	8m	4 bolts	
'Puits Colloque'			
Traverse	25m	Natural	Traverse above P80
		6 bolts	'l'Indomptable'
P12	15m	Natural	
P12	15m	2 bolts	
P8 and P34	50m	5 bolts	
P8	10m	2 bolts	

ANTRE DES DAMNES

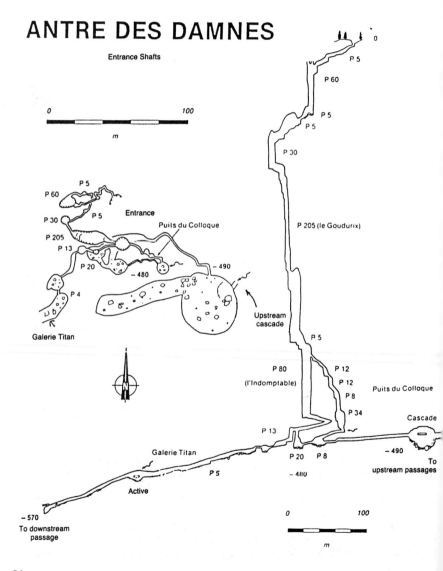

Entrance Shafts

SCIALET DE LA COMBE DE FER

Map: Carte IGN 1/25,000 No 3236 OT Villard de Lans/Mont Aiguille/Parc Naturel (TOP 25).

Map Reference: X: 851,80   Y: 304,67   Z: 1555

Depth: −582m
Length: 3,400m

Access
Leave Correncon in the direction of Clos de la Balme ski resort on the D215. Some 600m after the first hairpin bend there is a car park. On the right there is a small hut below an old teleski. Take the well marked path to the right. This climbs up into the Combe du Souillet for 3kms. (Ignore a left and right turn). Keep to the well trodden path until you reach a split in the path. Go left to climb up into the Combe de Fer in a south easterly direction where arrows are found marking the way across the lapiaz. Soon a small hut is reached. The entrance to the cave is in the large depression by the side of the hut at the foot of the steep slope that descends from the Tete des Chaudieres. It takes about 1¼ hours to walk to the cave from the car.

History
The large entrance passage was explored by E.A. Martel to −90m. In 1937 Bourgin and the S.C.A. Paris reached −183m at the base of P55. Despite some attempts during the 1950's (mainly by the Tritons) −183m remained the lowest point until 1963 when the A.S.V.(Villard de Lans), a group from the Grenoble Centre of Nuclear Studies and a team from the Youth Hostel forced a way through the 70m meander, stopping at −363m at the top of the 'Puits de la Boue'. The A.S.V. reached the 'Puits du Corail' in 1965 at −440m and the terminal siphon at −580m as well as surveying the cave. In June 1966 the same group found the 'Reseau de Juin' joining the 'Reseau Principal' in 1967 and continued their explorations until 1973 ('Reseau Nord' and the 'Reseau du Sommeil').

During the years 1967 and 1974/75 Belgian Cavers found the 'Reseau de Sissey' (or the 'Reseau des Belges') leading to a junction with the 'Puits de la Boue'. Since 1980 little has been found or documented except for a passage being found after the 'Salle Jesus'

Description
The 'Reseau Principal' to −580m via 'Puits de la Boue'. The large entrance passage descends to −90m over moraine where the passage becomes horizontal. (Beyond this passage P6 descends to the 'Reseau de Juin'). At the end of the passage several metres further on P10 gives access to the 'Reseau Principal'. From the bottom of the pitch the 'Puits Goupette' is found. This is a fine 35m pitch. Continuing from the base of this pitch a tight 2m gives access to P10 and then onto P55, (wet at the bottom). Do not take the passage that leads off from the bottom of the pitch (this ends after 200m), but climb up into the tiny passage 4m above. This enters the 70m meander, not particularly tight but very rough. This ends at P15, dropping into the 'Salle du Casse-croute'. Climb up a little to find another P15, dropping down to −213m. (From the bottom in the lower left hand corner the 'Reseau des Belges' is entered by a squeeze that leads into a meander that has a number of climbs down. A fine P20 leads into a further meander. After two climbs a

'toboggan' ends at P15 followed by a larger passage descending two small climbs of 4m and 3m. P35 joins the 'Reseau Principal' at the bottom of the 'Puits de la Boue'. The main way continues above and enters the 100m meander, starting with a 3m climb. This is drier and tighter but smoother than the 70m meander! A fine 45m pitch follows. At the bottom the meander severely shrinks for 3m and gives access to P60, a very spectacular pitch dropping into the 'Grandes Salles'. A passage going off from the lowest point descends to Camp 1 which leads to P25, the 'Puits de la Boue', at the convergence of the three important series of the Combe de la Fer.

Downstream P6 rejoins the streamway at −395m. A 10m cascade drops through a hole into a small chamber. 300m further on the siphon is met at −430m. This is passed by a very muddy 40m flat out crawl and a duck! Once the passage enlarges the 'Puits du Corail' (25m) gives access to an active and gloomy series of passages. To reach the following P15 one goes through a 3m long letter box where it is impossible to avoid the water. From the base of this pitch you need to crawl to reach P30. A pendule at −20m enters a tight rift and then P10 and a fossil passage. Another P10 arrives at the bottom of the 'Salle du Macaroni'. The next part consists of meanders and complex passages full of boulders until you reach the 'Salle du Bazar'. Beyond this point there are three climbs of 5m, 6m and 5m. From the bottom of the last climb the stream emerges from boulders and 150m further on the terminal siphon is reached at −580m.

Equipment

Shaft	Rope	Fixings	Remarks
P10	18m	4 bolts, Natural	Deviation at −4m
P35 Puits Goupette .	50m	3 bolts, 1 piton	Wet. Deviation at −30m
P10	15m	2 bolts	Tight at start, use traverse line
P55	65m	3 bolts	
P4	6m	1 bolt	Climb up to 70m meander
P15	18m	3 bolts	Deviation. Salle du Casse-croute)
P15	20m	3 bolts	−213m Salle du Carrefour
P3	4m	2 bolts	Start of 100m meander
P45	55m	5 bolts	
P60	70m	5 bolts	Arrive in Grande Salle
P25	30m	Natural, 2 bolts	Puits de la Boue
P6	7m	3 bolts	
P10	20m	4 bolts	Pendule to avoid water
P25 Puits du Corail .	27m	4 bolts	
P15	20m	3 bolts	
P30	35m	4 bolts	Pendule at −20m
P10	15m	2 bolts	
P10	15m	2 bolts	Salle du Macaroni
P5	10m	2 bolts	
P6	10m	2 bolts	
P5	10m	2 bolts	

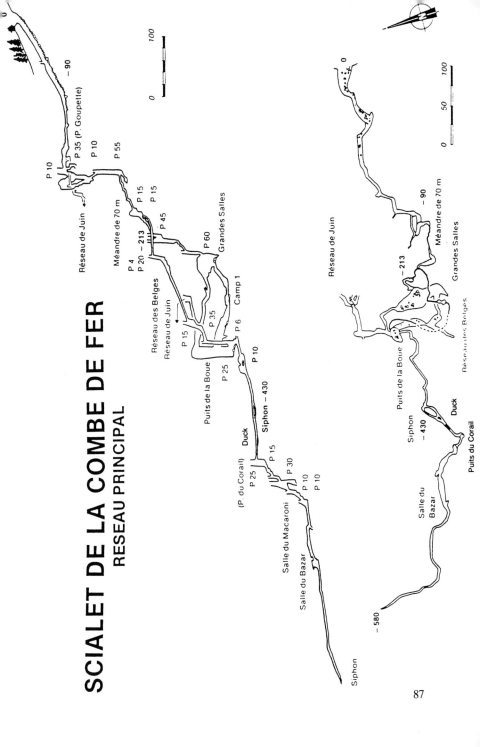

SCIALET DE LA COMBE DE FER
RESEAU PRINCIPAL

0 100

P 10
P 35 (P. Goupette)
P 10
P 55
Réseau de Juin
Méandre de 70 m
P 15
P 15
P 45
P 4
P 20 – 213
P 60
Grandes Salles
Réseau des Belges
Réseau de Juin
Camp 1
P 15
P 35
P 6
Puits de la Boue
P 25
P 10
Duck
Siphon – 430
(P. du Corail)
P 25
P 15
P 30
Salle du Macaroni
P 10
P 10
Salle du Bazar
Siphon

0   50   100

0
– 90
Réseau de Juin
Méandre de 70 m
– 213
Grandes Salles
Réseau des Belges
Puits de la Boue
Siphon
– 430
Duck
Salle du Bazar
Puits du Corail
– 580

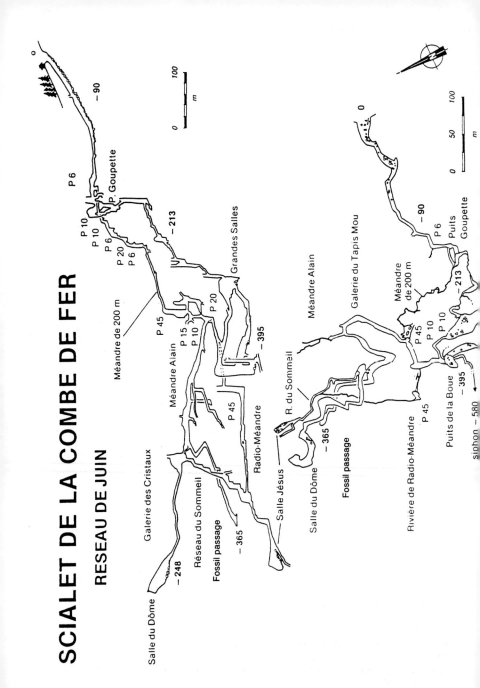

SCIALET DE LA COMBE DE FER

RESEAU DE JUIN

GROTTE DES DEUX SOEURS

Map: Carte IGN 1/25,000 No 3236 OT Villard de Lans/Mont Aiguille/Parc Naturel (TOP 25). The cave is marked on the map.

Map Reference: X: 855,74   Y: 305,55   Z: 1840

Depth: −315m
Length: 3250m

Access
The cave is situated at the foot of the Col des Deux Soeurs. There are two approaches. The first one is the most expensive, as it takes the Telepherique up to the 2000m station (which is actually only 1720m) from Le Balcon de Villard. From the station follow the path that leads to La Grande Moucherolle. When the path to the Moucherolle turns right keep straight on and over the Col. Descending from the Col find the hole to the left on a ledge just above the first rock band (last going down) of the cliff.

The second and perhaps the best approach is from Prelenfrey, Commune du Gua. Take the N75 from Grenoble as far as Vif then the D8 to Prelenfrey. From the village take the forestry road to a car park at point 1309. After several hundred metres there is a junction of paths. Take the right hand one, marked Pierre des Deux Soeurs. This path is the 'Sentier Balcon-Est' which traverses underneath the main spine of the Vercors. Follow this path to the well built refuge of Clos. 200m further on follow a zig-zag path on the right up scree. At the next junction go left to the foot of the cliff to where the hole is found on the right of the path above the first rock band. This way takes about 1½ hours.

History
That cave was discovered in 1902 by Fonne and Muller and descended to −40m. In 1944 A. Bourgin and the Clan E.D.F.Lesdiguieres reached −90m but were stopped by a squeeze. During 1951-52 the S.G.C.A.F. passed this point and reached −145m in the 'Reseau des Grenoblois'. The following year the same team along with members from Les Esclaireurs de France de Lyon (Verna-Tritons) explored the 'Reseau des Enrages' and also found the 'Puits de la Verna' and the 'Salle du Lion'. In 1954 the siphon was reached at −315m. In 1978 the Speleo Club de Vizille found some climbs in the 'Reseau des Grenoblois'.

Description
The stony and earthy entrance draughts strongly and enters a sloping passage which has three short climbs, 3m, 3m and 5m to reach −40m where the start of the 'Reseau Lesguidieres', ascending to +17m, is found. The main way continues past two short, close together pitches (P6 and P9) to reach −90m where a very muddy 30m crawl is entered. Two other short crawls and meanders reach the 'Salle Dominique' and then the 'Salle de Douche'. From here three ways lead off, the 'Reseaux Enrages' and 'Grenoblois' and the 'Reseau Verna'. Below the 'Puits de Douche' a stony passage reaches the 'Puits de la Vire' where, above, a stone shoot gives access to the 'Reseau Enrages'. A ledge above the 'Puits de la Vire' gives access to the 'Reseau des Grenoblois' which is characterised by low passages and shafts that disappear upwards.

The 'Reseau Verna' starts with the 'Puits de la Verna' (70m) quickly followed by P15. A short meander reaches another P15 giving access to a sloping passage at −190m. This ends at the 'Puits du Lion' (50m). At the bottom of this pitch the passage becomes wet but after 50m a dry passage bypasses the siphon at −225m and rejoins the stream further on by a 6m climb. After 50m another dry passage has two descending branches (P10 or P15) to reach the siphon at −315m.

Equipment
To reach −315m

Shaft	Rope	Fixings
P5	8m	2 bolts, 1 piton
P6/P9	20m	4 bolts
P70 (Puits de la Verna)..........	85m	3 bolts, 1 deviation
P15	20m	3 bolts
P15	20m	3 bolts
P50 (Puits du Lion]	60m	3m sling and 4 bolts
P6	10m	2 bolts
P10	15m	2 bolts
or		
P15	20m	2 bolts

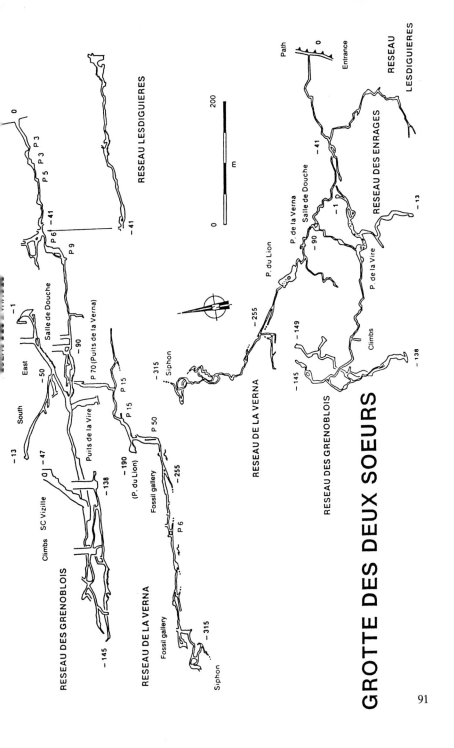

GROTTE DES DEUX SOEURS

91

SCIALET DE MALATERRE

Map: Carte IGN 1/25,000 No 3236 OT Villard de Lans/Mont Aiguille/Parc Naturel (TOP 25). The cave is marked on the map.

Map Reference: X: 848,62 Y: 309,56 Z: 1418

Depth: −230m
Length: 1600m

Access
From Villard de Lans follow the road, D215c, towards Herbouilly. Take the second turning off to the left 2 kilometres beyond Bois Barbu on a right hand bend. 400 metres later turn right where the road goes straight on (Route Forestiere de Combe Verte) and drive along here for 3·5 kilometres. 200 metres after a turning to the right there is parking space on the side of the road, signposted to the Scialet. A 5 minute walk takes you to the bridge spanning the shaft.

History
Although the hole has always been known it was not explored until 1936 by Andre Bourgin and the C.A.F. de Paris who were stopped at −159m.
  The large chamber at the base of the cave at −230m was reached in 1971 by the Speleogroupe des Hauts de Seine. In May 1979 the S.G.C.A.F. discovered a shaft, parallel to the main one, which started at −50m and entered the main shaft at −110m.

Description
A magnificent and airy entrance shaft can be a real trouser filler for some! Taking out a panel from the floor of the bridge one can rig a free hang for the first stage to a sloping ledge at −55m. (From the re-belay here it is possible to enter, by climbing easily up to the right, the parallel shafts P10, P8 and P22 to a meander that leads onto P10 and down to the base of the main shaft). Care has to be taken on this ledge not to disturb any loose stones on to those below. Two short sections lead to the final part of the shaft. (Re-belay at −105m). The landing at −120m is exposed to stonefall so take shelter away from the line of fire! The view up is remarkable. It feels as though a pair of eyes are looking down on you!!
  A short climb down a loose rubble slope leads to a low, but wide and steep passage with a number of routes leading off. The way on is a narrow meandering passage leading quickly to P16 and dropping down to the 'Ancien Terminus' at a depth of −159m.
  Above P16 a traverse in a rift leads to a draughting low crawl before reaching a fine shaft of 30 metres. The key hole shaped passage one enters is the 'Galerie de Meandre'. There are several ways underneath the meanders, which are very muddy and wet. It is not far to P20. The continuation is tedious and can only be done in dry weather. Follow the ensuing fairly high rift to the right. The way is very muddy and slippery and soon reaches an area of partial siphons. These give access to the 'Grande Salle' and the base of the cave at −230m. Diverse other passages exist in this area. Only the dedicated with a desire to reach the bottom will continue beyond the base of the main shaft.

Equipment

Shaft	Rope	Fixings
P120	135m	1 sling, 6 bolts
P16	25m	4 bolts
P30	37m	5 bolts
P20	25m	4 bolts

Ensure that the rope does not rub on the bridge.

If going to −230m take 4 extra ropes of 15m for climbs and awkward passages.

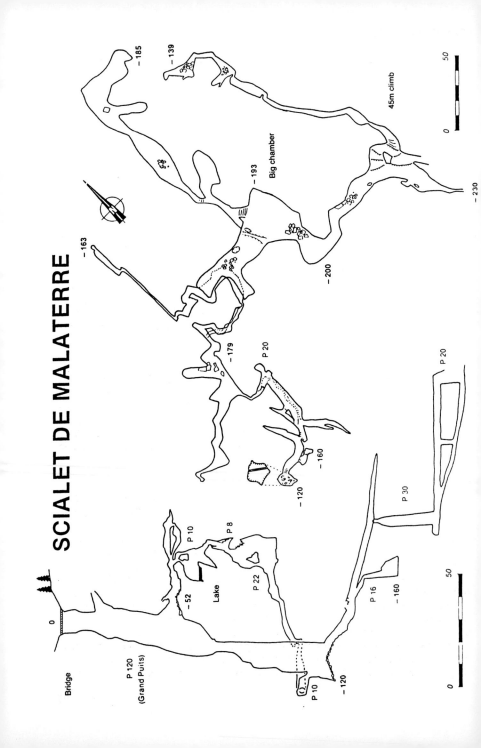

SCIALET DE MALATERRE

Bridge

0

P 120
(Grand Puits)

−52

Lake

P 10

P 8

P 22

−120

P 10

P 16

−160

P 30

P 20

−120

−160

P 20

−179

−163

−185

−139

−193

Big chamber

−200

−230

45m climb

0   50

0   50

SCIALET DU TRISOU

Map: Carte IGN 1/25,000 No 3236 OT Villard de Lans/Mont Aiguille/Parc Naturel (TOP 25).

Map References: Scialet du Trisou   X: 847,28   Y: 309,66   Z: 1368   Scialet du Regard
X: 847,31   Y: 309,62   Z: 1370

Depth: −273m
Length: 1,388m

Access
From Villard de Lans take the road, D215c, as if going to the Scialet de Malaterre. Instead of turning left, keep to the right and follow the Chemin de Croix road through the forest past the Belvedere de Valchevriere, heading for the Herbouilly pasture to the Chalet de Chalimont. Continue for another 1·2 kilometres to where a large doline is easily seen to the left a few metres from the road. Parking is difficult here but with care cars can be parked off the road. A path descends into the doline. Another doline has the Scialet du Regard but the Scialet du Trisou is the larger of the two entrances.

Approaching from Saint Julien-en-Vercors follow the road to Saint Martin-en-Vercors and midway between the two villages turn left on the D221 as if going to the Grotte des Ramats but continue to the Herbouilly pasture. Keep following the road towards Villard de Lans and as the road descends into the wood the doline is seen to the right, 150m beyond a forestry track on the right.

History
The cave was explored for the first time in 1937 by the G.S.A.P. (Bourgin and Glory) when they descended to −84m, stopping at the 'Chatiere de l'Abbe Mouton'. Twenty years later the cave was descended to −240m (the 'Puits de l'Infini') by the G.S.V. and the Tritons. One year later on the 12th October 1958 the siphon was reached at −273m. In July 1961 a group of C.E.N.G. discovered the Scialet du Regard and a month later reached the junction with the stream passage at −72m. M. Chiron explored upstream in 1978.

Description
Descend the doline slope carefully to the cave entrance and P3, shortly followed by P18, a fine free hang. Follow steeply sloping meanders to the head of P21. Traversing round the corner reveals bolts for a free hang to a ledge 6m from the floor where the streamway is reached. Continue downstream, climbing over stal blockages to reach a short crawl, 'Chatiere de l'Abbe Mouton'. A short traverse leads to P12, dropping into a small chamber. The way out is an interesting P3 climb down called the 'Rapide'! Traversing slippery ledges surmounts the next barrier from where P18 descends into a rift. A short walk leads to the next step of 2m, followed immediately by P37 the 'Puits de la Douche'. The serious caving starts here as all the successive pitches can be impassable when the cave starts to flood! A tide mark half way down the pitch suggests that the whole of the lower cave fills up. To keep dry a long bolt traverse is taken to keep out of the main force of the water. Avoid the temptation to stop short of the end of the traverse – it may be a lot wetter on the way out! (It is possible to avoid the P18 and P37 by taking a slippery 30m traverse above the 'Puits de la Douche' starting shortly after the climb

down by a stal flow. A rope of 40m is recommended here. (Although the P56 pitch is longer it is much drier). A meander at the bottom of the pitch is gloomy and foreboding with spray in the air. The stream is followed to P21, 'Puits des Djinns', then onto a double cascade, P12. Thirty metres further on 'Puits du Sac' (P29) is reached. The cave is now even wetter and more foreboding in the black rock. After some small cascades an 11m chute is descended and then at −227m the 'Puits de l'Infini' (P46) is entered by a 2m diameter window. Rock hereabouts bristles with chert and fixings are found to the left on the roof. Go down 6m to reach a sheltered rebelay. The remaining drop is very wet. A rift is followed to a cascade blocking the passage. Traversing this the siphon is reached after a few metres at −273m.

Equipment

Shaft	Rope	Fixings	Remarks
Doline	—	Tree	Hand line
P3	—	2 bolts	
P18	45m	2 bolts	
P21	30m	4 bolts	
P12	15m	sling	
		2 bolts	
P3	5m	2 bolts	
Traverse	40m	4 bolts	Slippery
P56	65m	3 slings	
		2 bolts	Wet
P21	30m	4 bolts	Wet
P12	25m	4 bolts	Wet
P29	40m	sling	
		2 bolts	Wet
P6	—	2 slings	Wet
P1	—	2 slings	Wet
P4		2 slings	Very wet
(Puits de l'Infini)	80m	3 bolts	Bolts to left on roof

SCIALET DU TRISOU

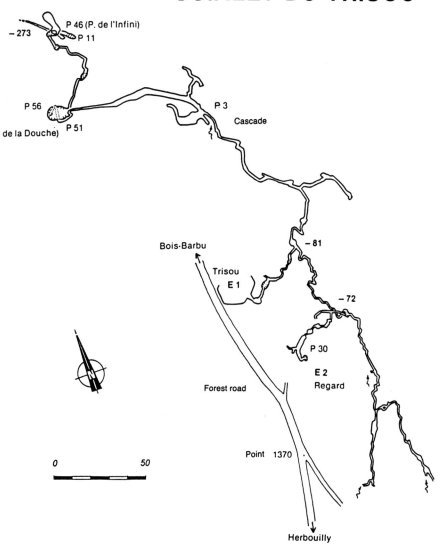

P 46 (P. de l'Infini)
P 11
− 273

P 56
P 51
de la Douche)

P 3
Cascade

Bois-Barbu

Trisou
E 1

− 81

− 72

P 30

E 2
Regard

Forest road

0   50

Point   1370

Herbouilly

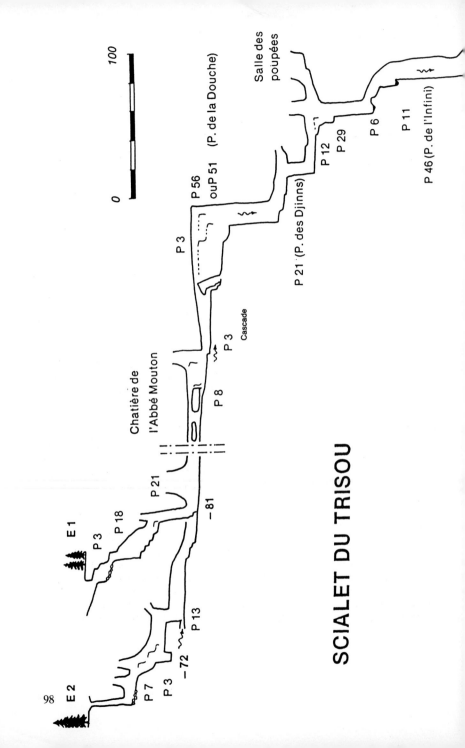

SCIALET DU TRISOU

SHOW CAVES

There are a number of show caves in the area ideal for an off day or for taking the family. They are all reasonably priced as well as being spectacular. Below is a resume of them and how to get there.

Grottes 'Les Cuves' de Sassenage
These caves are in Sassenage and well signposted once in the village. If staying at Autrans, Pont en Royans or Villard de Lans, take the N531 out of Villard de Lans in the Grenoble direction, drive through Lans en Vercors and Engins to Sassenage. This cave is the bottom end of the Gouffre Berger, with water taking 48 hours to find its way through from the terminal siphon. The cave is only open from 1st May to 31st October.

Grottes de Choranche
The manager of these world famous caves must be asked for permission to explore the Grotte de Gournier. Approach as for that cave, from the well signposted road that runs through the Bourne Gorge. They are open all year and have a marvellous array of straw stalactites, some over 3m long, as well as other fine formations in the Grotte de Couffin. The Couffin has recently been extended to include the huge Cathedral chamber of the Grotte de Chevaline. It's formations are lit up in a spot lighted sequence arranged to dramatic music!

Grotte de la Luire
Take the road from La-Chapelle-en-Vercors towards St-Agnan-en-Vercors (D518). From here follow signs to the car park. A short walk takes you to the entrance. The cave is open from April to October. Used as a hospital by the Resistance during World War 2 the cave is one of the hydrological wonders of France.

Grotte de la Draye Blanche
From La-Chapelle-en-Vercors take the road towards Vassieux (D178). The cave is found on the right hand side of the road 1·5kms past the turning to the Col de Carri. This is the highest show cave in France and is open from April to October. It has a fine array of columns and a huge stalagmite flow with different colours, from snow white to dark brown.

Grotte de Thais
This cave is set in the middle of the village of St-Nazaire-en-Vercors at the base of the aqueduct. From Pont en Royans take the D531 through Aubrives to join up with the N532. Turn left to St-Nazaire-en-Vercors. The river in the cave is one of the most important in the Vercors as well as having most unusual red stalagmites. There is an exhibition of cave life with aquariums and vivariums, and the lifestory of prehistoric people who lived here 15,000 years ago. The cave is open every day from 1st June to 30th September and during April, May and October it is open on Sundays and holidays.

SELECTED BIBLIOGRAPHY

All books and magazines are written in French unless stated otherwise.

SPELEO SPORTIVE DANS LE VERCORS. 1987, by Jean-Jacques Delannoy and Dominique Haffner. Published by Edisud. A selected cave guide to the region.

PAYSAGES DU VERCORS SOUTERRAIN. 1981, by Jean-Jacques Delannoy and Baudouin Lismonde published by Le Comite Departmental de Speleologie de l'Isere. Describes the area in photographs with a commentary and pictures of some of the caves, Grotte de Gournier, Grotte de la Luire, Grotte de Bury, Reseau Christian Gathier, Reseau Couffin – Chevaline (show cave), Cuves de Sassenage (show cave), Scialet de la Combe de Fer and of course the Gouffre Berger, among others.

GROTTES ET SCIALETS DU VERCORS. 1985, – THE definitive guide to the area in two volumes but as one large tome! Again published by Le Comite Departmental de Speleologie de l'Isere and written by Baudouin Lismonde and Jean Michel Frachet amongst others.

GOUFFRE BERGER, PREMIER 1000. 1977, by Georges Marry. A black and white photographic essay of this marvellous cave.

ONE THOUSAND METRES DOWN. 1957, by Jean Cadoux and others. Published in English by A.S.Barnes and Co. THE story about the exploration of the Gouffre Berger.

LE TROU QUI SOUFFLE. 1992, by Baudouin Lismonde. Published by Le Comite Departmental de Speleologie de l'Isere. The story of all the explorations of this remarkable cave with fine photographs.

To keep up to date with recent discoveries 'SCIALET' published annually by Le Comite Departmental de Speleologie de l'Isere is recommended. Current and back copies can be bought at EXPE.